Health and Safety

. . . in brief

Fourth Edition

John Ridley

ELSEVIER

Amsterdam • Boston • Heidelberg • London • New York • Oxford
Paris • San Diego • San Francisco • Singapore • Sydney • Tokyo

Butterworth-Heinemann is an imprint of Elsevier

Butterworth-Heinemann is an imprint of Elsevier
Linacre House, Jordan Hill, Oxford OX2 8DP, UK
30 Corporate Drive, Suite 400, Burlington, MA 01803, USA

First published 1998
Revised reprint 1998
Reprinted 1999, 2001
Second edition 2001
Third edition 2004
Fourth edition 2008

British Library Cataloguing-in-Publication Data
A catalogue record for this book is available from the British Library.

Library of Congress Cataloging-in-Publication Data
A catalogue record for this book is available from the Library of Congress

ISBN: 978-0-7506-8639-6

For information on all Butterworth-Heinemann publications visit our web site at books.elsevier.com

Printed in Great Britain

08 09 10 11 12   10 9 8 7 6 5 4 3 2 1

Contents

Preface

The text of this book started life as a set of notes I used for lectures on health and safety which became popular with the students as *aides-mémoire* for their revision prior to examinations. The students came from various positions in commerce and industry, ranging from union appointed safety representatives through middle managers to directors. Feedback from them indicated that they found the notes useful in their day-to-day duties. If they could, why couldn't others – and so the book was born.

In preparing this book I have endeavoured to take some of the mystique out of health and safety in the workplace and in simple plain English and in practical terms, explain the hows, whys and wherefores of keeping on the right side of health and safety laws.

It is not intended to be a substitute for the law. Readers must understand that when the crunch comes it is the responsibility of the employer/manager to ensure compliance with legislative requirements and, in the ultimate, for the courts to decide if compliance has been achieved. It is hoped that this book will help those responsible to meet their obligations and avoid running foul of the law.

There can be no doubt, the subject is vast and complex and to meet the objective of simplifying health and safety laws, their contents have been polarized into major components and the overall subject broken down into manageable parts. The text is purposefully kept brief – making use of 'bullet points' rather than verbiage – with references to sources of information should a particular situation demand more detailed study.

The text is restricted mainly to matters that are likely to be met in the day-to-day work of a busy manager with safety responsibilities, an emergent safety adviser, safety representative or those participating in safety awareness courses. While it may be of practical help in the design and manufacture of equipment, designers and manufacturers should refer to the various legislation and standards that relate to their particular product. This latter point is especially important within the context of the European Union.

With the written word it is not possible to include the inflexions of speech or the communication of body language so a convention has been adopted to differentiate between what is law and what is advice:

titles of and quotations from legislation are in	*italics*
extracts from laws are in	boxes with shading and indicated by a symbol in the left hand margin
advice and comments are in	brackets or normal type face.

I have also used abbreviations when referring to specific sections of an Act as 's. – ' and to the clauses of a Regulation as 'r. – '

The advice and comments given draw on my experiences through the years of involvement in the drafting, enforcing and complying with various aspects of health and safety legislation. They are not intended to be comprehensive or exclusive but aim to draw attention to some of the essential features of the law and of current practices so that the individual, be he manager or safety adviser, can with reasonable confidence, and in the culture of his particular organization, make decisions on how to proceed or what action to take. Nor are they intended to denigrate legislative requirements, but to explain them in a simple manner so the task of complying is made easier for those working at the 'sharp end' of industry and commerce. Where a particular problem is met, reference should be made to the relevant legislation and published guidance on the subject.

The book is in four parts dealing with:

– the background to health and safety law
– employer's (and manager's) role and responsibilities
– health aspects
– technical aspects

Specific laws with health and safety connotations are dealt with where they arise in the context of the particular matter being dealt with, rather than including them in a general law section.

As far as possible the text has been written in the impersonal, but in places it has not been possible to do this so I have referred to *he, him* and *his*. This is not sexist but for convenience and I fully recognize and acknowledge the part that women play at all levels in our work lives. So wherever *he, him* or *his* occur please also read *she, her* or *hers*. I hope female readers will forgive me this convenience.

The contents cover in outline the syllabus for the National General Certificate level examination of the National Examination Board in Occupational Safety and Health and of the requirements for the Working Safely and Managing Safely Certificates sponsored by the Institution of Occupational Safety and Health. It should also prove a useful *aide-mémoire* for those studying for higher level health and safety examinations.

To save repeating long titles, especially where they occur frequently in the text, I have used initials and have listed them in an appendix. There is also a comprehensive index to enable particular subject matter to be located easily and eliminate the time wasting task of wading through pages of text to find the pertinent information.

Health and safety laws are not static but are developing all the time to meet changing conditions and technologies. This situation is exacerbated by our committment to membership of the European Union which is increasingly dictating the subject matter of our laws.

The subject range of this book is vast and I have been grateful for the help, assistance and encouragement given to me in its preparation by Robin Rispin and Ray Chalklen. I would also like to thank Stocksigns for their help in providing copies of the hazard signs and John Stather of NRPB for his comments on Table 9.1.

Health and safety is not a serious subject in the sense that it is *dreary*, but it is serious in the sense that it is *important*, and as such is something that can be enjoyed and laughed at. So, to leaven the text, I have included a number of cartoons which I hope you will enjoy but which will also help you to remember particular points. I am grateful to Philip Wilson for those cartoons and showing us a lighter side to some of the stock safety phrases we take so much for granted.

In a book of this sort it is only possible to summarize very briefly the nub of legislative requirements, but I hope I have included sufficient information to give managers the basic information they need and safety advisers and students the trigger to remind them of more detailed requirements. One of the aims of the book has been to simplify the legislative requirements that are wrapped up in complex Parliamentary verbiage and it has done this by hacking away that verbiage to expose the basic objectives of health and safety laws in readable everyday English. I hope I have succeeded.

John Ridley
October 1997

Note to the second edition

Since the first edition was published, there have been a surprising number of developments in, and rationalization of, legislative requirements. There have also been changes in the way in which EU directives are adopted as a result of the co-decision procedure introduced by the Maastricht and Amsterdam Treaties. This has given the European Parliament a much greater say in the content of directives.

In the UK, as well as the on-going development of the COSHH and CHIP Regulations, new regulations have been introduced for lifting operations, press brakes and woodworking machinery. In addition, a considerable number of new booklets of advice and guidance have been published.

I have endeavoured to include all these changes without making too many alterations to the original text.

John Ridley
January 2001

Note to the third edition

The need for a new edition has been precipitated by a spate of directives from the EU dealing with hazardous substances, asbestos and lead requiring a revision of the existing relevant regulations. In addition, a new regulation, The Dangerous Substances and Explosive Atmospheres Regulations 2002 has come into effect superseding the High Flammable Liquids and Liquified Petroleum Gases Regulations 1972. Also, The Working Time Regulations 1998, which has implications for health and safety, has been receiving much attention.

Major changes are occurring in the legal system and its procedures as a result of the Woolf Report and the Auld Report. While the recommendations of those reports have been accepted and are being implemented, inevitably, things legal take time to change but changes are in train and are likely to accelerate as their benefits become apparent.

An event, in August last year, that brought the major players in the health and safety field together to join with the Employment National Training Organization, was the launch of the *New Standards in Occupational Health and Safety Practice*. These new standards define the requirements for the various levels of National Vocational Qualifications (NVQs) in Occupational Safety and Health, qualifications that can contribute towards membership of the Institution. This updated text covers the range of knowledge required for those NVQs.

I have endeavoured to include all these changes with the minimum of alteration to the existing text.

John Ridley
November 2003

Note to the fourth edition

Health and safety legislation continues to flourish. Since the third edition there have been a number of major developments, some reflecting EU directives and others rationalizing domestic laws. Areas affected include chemicals through the burgeoning REACH Regulations. Vibration regulations have been introduced as a logical adjunct to the new noise regulations. In construction many of the extant disparate requirements have been brought under one roof by the CDM Regulations 2007, which is complemented by the Work at Height Regulations, whose application extends beyond the construction industry.

The requirements for fire safety have been revamped by the Regulatory Reform (Fire Safety) Order 2005, which seems to owe much of its substance to the Health and Safety at Work Act 1974. It removes from the Fire Authority the onus of issuing Fire Certificates and puts the responsibility for fire safety firmly in the court of the employer. Finally, requirements in respect of the handling and disposal of hazardous wastes have been extended and tightened.

One non-legislative area where developments are occurring is as a reaction to the misuse of risk assessments that has sired a risk aversion culture. There is official recognition, both in law and by the HSC, that some activities necessarily carry risks, but where risks do occur they must be properly managed.

In this edition, I have endeavoured to include all these points and to bring the text as up-to-date as possible.

John Ridley
September 2007

Part 1   Law

The system of English justice was not designed but developed gradually over more than 1000 years. It is a system that has been used as a model in many parts of the world.

In parallel with the development of the system so the method of making laws has evolved and now follows a well proved procedure although there have been some radical changes in the past two or three decades.

Since 1972 a major influence on the subject matter of English laws has been the European Economic Community (now the European Union) where, as a Member State, the UK is required to incorporate into domestic laws the content of EU directives once they have been adopted by the Council of Ministers.

This part deals with the historical background to, procedures for making, administering and enforcing health and safety laws as well as covering the content of two of the main pieces of health and safety legislation.

1 Legal processes

The English legal system and its laws did not happen over night, they took thousands of years to develop. During that time they were adjusted and modified as the system of justice became more consistent across the nation. For the last 700 years there have been only relatively minor developments although the past two decades have seen a number of important changes in the role of members of the legal profession and, more recently, in the procedures for dealing with cases.

This section looks at the background to law in the UK, how laws are made and the effects of the increasing influence of the European Union on UK laws.

1.1 Historical background

After the Romans left England around AD 400, the country split into a number of small kingdoms each ruled by the local lord. This gave rise to great injustices and misuse by the lords of their powers. This situation continued until the latter part of the ninth century when the gradual unification of the country began. The English system of justice was not designed; it developed slowly over more than 1000 years through moves to correct particular problems and injustices, and it is still developing today. Some of the major steps in that development are summarized below.

Alfred the Great	871–901	• unified England • drew up a Book of Dooms or Laws • introduced an administrative structure for government
Edgar	959–975	• set up Shire, Hundred and Borough courts to administer the Dooms
Edward the Confessor	1042–1066	• first effective king • introduced King's Justices to administer more consistent justice across the country
William the Conqueror	1066–1087	• rationalized the existing legal arrangements • developed the English Constitution • established the 'Opposition'

		• established his own 'superior' courts to get away from the parochial nature of existing courts • initiated the doctrine of legal precedents
Stephen	1135–1154	• moved administration of the country away from the Crown • birth of the 'Civil Service'

Henry II	1154–1189	• created King's Courts to try breaches of the King's Peace
		• introduced trial by jury
		• introduced the concept of Royal Writs to attract litigants to the King's Courts and give them fairer justice
		• administered the courts through a Royal Council:
		  – Court of Exchequer
		  – King's Court or Bench
		  – Court of Common Pleas
		• appointed travelling (Circuit) judges to sit in the King's Courts around the country
		• set up the framework of law that is followed today

John	1199–1216	• signed Magna Carta in 1215 giving many law-making powers to the lords and barons
Henry III	1216–1272	• established first Parliament of elected representatives of landowners and citizens

Edward I	1272–1307	• Edward the Law Giver
		• made House of Lords the chief Appeal Court
		• established three departments of the King's administration:
		  – the Exchequer
		  – the Chancery
		  – the Wardrobe
		• specialist professional legal groups became established, i.e. the Inns of Court: Clements, Lincolns, etc.
		• passed statute declaring that English legal memory started when Richard I came to the throne in 1189 as a tribute to the efforts of Henry II in establishing a legal system

From this time, with some small developments, the legal system that existed at Edward I's death has been followed to the present day.

However, in the last two decades or so there have been developments in the administrative arrangements within the legal system such as the setting up of Employment Tribunals and the giving to solicitors the right to appear in the higher courts and to become judges. Also, following the recommendations of the Woolf and Ault reports, changes are being implemented to shorten the procedures for dealing with cases.

1.2 **Branches of law**	English law divides into two major categories:

1   **Statute law** comprises those laws which are debated and made by Parliament. They are written and copies are available for all to purchase (from The Stationery Office (TSO)). Statute law is the law of the State and a breach of it is a criminal offence, hence it is sometimes referred to as criminal law.

2   **Common law** originally referred to a system of law common across the whole country. However, its meaning has changed to encompass **case law** – the body of law based on **judicial precedents**, i.e. the decisions of judges in earlier cases are binding in similar later cases. It is not *officially* written but is recorded in Law Reports.

Equity law supplements common law in that it deals with complaints against common law judgements where there may have been a distortion of justice, i.e. such as from undue influence by one party.

There is some overlap between these categories, for example, statute law can rely on interpretations from common law and common law cases can rely on breach of statutory duty to support a claim.

Statute law divides into:

1   **Public law** – dealing with matters involving the State and relating to the protection and well being of the public at large whether directly or indirectly. Typical of these laws are:

- constitutional laws
- employment laws
- social security laws
- criminal laws

2   **Private law** – statutory requirements that deal with matters regulating the relationship between 'private parties' or 'legal and natural bodies' such as in:

- contracts
- torts
- property

- trusts
- succession

3 **Civil law** – a generic term covering all laws except criminal laws.

1.2.1
Legal actions

Criminal

- Law concerning the protection of the State, the community and the individual.
- Laws are passed or approved by Parliament.
- Written law, copies obtainable from TSO.
- A breach is a crime or criminal offence.
- Criminal proceeding (prosecution) heard:
  - in Magistrate's Court for lesser cases
  - in Crown Court for more serious cases.
- Evidence must prove guilt 'beyond reasonable doubt'.

Civil

- Common law action.
- Between two or more private or corporate bodies.
- Based on precedent set by judgements in earlier cases.
- Proceeding (litigation) heard in County Court.

The meaning of words and phrases of statutes are often critical to the outcome of a case and a judge may define his understanding or decide the interpretation of a word or phrase when giving his judgement.

1.2.2
Interpretations

Decided cases – those cases in which, as part of a judgement, a word or phrase that is part of a law is defined, i.e. the meaning is **decided**. Once a meaning is decided, it applies to all laws. A good example is the phrase *so far as is reasonably practicable* which was defined in a civil case in 1949 concerning a mining accident [*Edwards v. National Coal Board (1949) 1 AllER 743*], long before the phrase was incorporated into the *Health and Safety at Work, etc. Act 1974*.

Tort – a wrongful act or omission causing harm or damage to a person or body corporate and which is actionable in common law. Typical torts include:

- nuisance
- trespass
- negligence[*]
- breach of statutory duty[*].

[*] usual causes of actions in health and safety

Rule of law

- No person or body is above the law
- There is one system of law for everybody.

Supremacy of Parliament

When Parliament was the originator and maker of the laws of the country, there was nothing that Parliament could not lawfully do. However, with the UK's commitment to the EU in 1972 Parliament's law-making powers became restricted in those areas covered by the dictates from the EU in the form of Directives and Regulations.

Figure 1.1 Sources of English law

1.3
Types of court

Matching the two types of law, i.e. statute and common, are two sets of court hierarchy, i.e. criminal and civil.

1.3.1
Criminal courts

(a) Magistrate's Court

- Dispenses summary justice, i.e. trial without a jury.
- Normally comprises a bench of between two and seven lay magistrates. Lay magistrates are not legally qualified and are unpaid. In court, they are advised on matters of law and procedure by the Clerk to the Court, who is a barrister or solicitor with at least five years experience.
- In some cities cases may be heard by a **stipendiary magistrate**, who sits alone but is a professionally qualified barrister or solicitor of at least seven years standing.
- Cases are heard in public.
- Deals with minor criminal cases.
- Has limited powers of sentence:
  - maximum fine of £20 000
  - prison sentences up to three months for cases involving breaches of the conditions of a Prohibition Notice.
- Hears preliminary evidence in certain serious cases before committing them for trial in the Crown Court.

(b) Crown Court	• Court of **first instance**, i.e. 'hears' cases for the first time, the first level at which a case is officially recorded.
	• Trial by jury.
	• Hears trials on indictment.
	• Both parties are represented by solicitor or barrister.
	• Hears appeals from a Magistrate's Court.
	• Will pass sentence on cases referred to it by a Magistrate's Court.
(c) Court of Appeal	• Hears appeals from the Crown Court.
(d) House of Lords	• Hears appeals from the Appeal Court.
	• The final appeal court in the UK.

1.3.2
Civil courts

(a) County Court	• Covers a restricted geographical area.
	• Hears a wide range of cases in the first instance.
	• Presided over by a Circuit Judge or a District Judge who has limited powers of sentence.
(b) Employment Tribunal	• Less formal than a court.
	• Sits under a Chairman who is a solicitor.
	• Chairman is supported by one representative appointed by an employer's organization and one appointed by an employee's organization.
	• Procedure to be followed in making an appeal is laid down in *The Employment Tribunals (Constitution and Rules Procedure) Regulation 1993* as amended.
	• Applications to appeal to a Tribunal must be made within a prescribed period depending on the cause of complaint.
	• Both appellant and defendant can represent themselves.
	• Hears appeals against:
	– Improvement Notices
	– Prohibition Notices
	– refusal by an employer to give a safety representative time off with pay to carry out his functions or to attend training in his functions.
	• Hears appeals on employment matters regarding:
	– redundancy
	– maternity leave
	– racial and sexual discrimination
	– unfair or constructive dismissal
	– equal pay.
	• Appeals against a Tribunal's decision may only be made on points of law.
(c) Employment Appeal Tribunal	• Associated with the High Court.
	• Sits under a judge with 2 to 4 lay members all with equal say.
	• Hears appeals from an Employment Tribunal on matters of law.

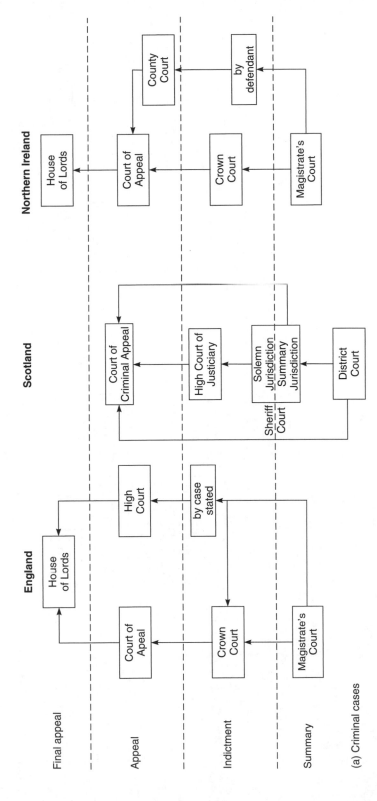

Figure 1.2 The different courts in the UK

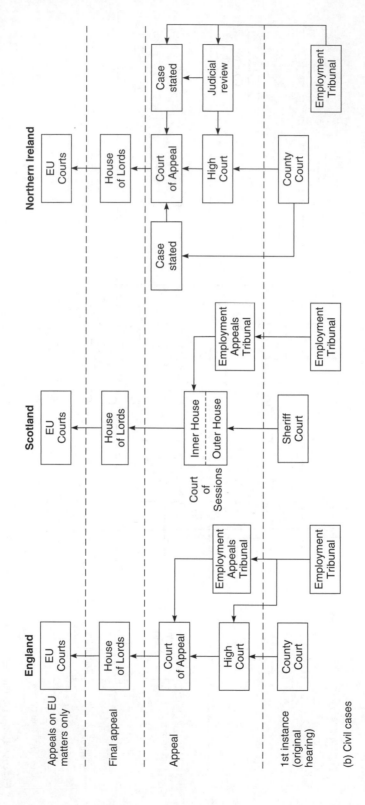

Figure 1.2 Continued

(d) High Court

- A group of three superior Courts having both first instance and appeal functions.
- Divisions of the High Court are:
  - Queen's Bench, dealing with damages, debts and property
  - Family, dealing with adoption and matrimonial matters
  - Chancery, dealing with companies, wills, bankruptcy and intellectual property.
- Divisions of the High Court hear appeals from:
  - Industrial Tribunals on matters of law
  - County Courts.

(e) Court of Appeal

- Appeals heard by three judges.
- Hears appeals from:
  - Employment Appeal Tribunal on matters of law
  - County Courts
  - Divisional Court.

(f) House of Lords

- The ultimate point of appeal in the UK.
- Appeals considered by three or more Law Lords.

(g) European Court of Justice

- Concerned only with matters relating to European Union legislation.
- Cannot interfere with justice in a member state.
- Judgements can require a member state to reconsider its own judgement.

Some of the courts and routes of appeal in Scotland and Northern Ireland are different from those in England. These are shown in Figure 1.2(a) and (b) for criminal and civil cases respectively.

**1.4
Court procedures**

Behaviour and procedures in court have developed over many hundreds of years and, in general, the procedure followed is very similar in all the courts, the main differences being the degree of formality and who is allowed to appear. Below are outlined the procedures followed in the first levels of court only, i.e. Magistrate's Court, County Court and Employment Tribunal, since it is unlikely that many readers will become involved in the complex procedures of appeals which are handled by professionally trained legal people.

Described below are the procedures for English courts. Certain of the procedures followed in the courts in Scotland and in Northern Ireland differ from those in English courts. Some indications of these differences are given in Figure 1.2.

1.4.1
Magistrate's Court

A court of summary jurisdiction in which criminal cases, i.e. alleged breaches of, *inter alia*, health and safety laws, are heard by a 'bench' of between two and seven magistrates although in some cities the cases may be heard by a stipendiary magistrate sitting alone. Stipendiary magistrates are full-time, are paid a stipend and are usually barristers or solicitors with at least seven years legal experience.

Representation in court:

- The prosecutions in criminal cases are conducted by either a solicitor or barrister, except for health and safety cases, where the HSE (Health and Safety Executive) Inspector is authorized to conduct his own prosecution.
- The defendant can represent himself but in health and safety cases it is usual for the defendant to be represented by a solicitor or barrister. If there is a common law claim pending it is likely that the defendant's EL (Employer's Liability) insurers will insist on, and pay for, professional representation.

The procedure is as follows:

- The prosecuting inspector presents his own case.
- The defendant is normally represented by a solicitor or barrister.
- The charge is read out by the Clerk of the Court.
- The defendant is asked how he wishes to plead, guilty or not guilty.

If he pleads **guilty**:

- The prosecutor summarizes the circumstances of the alleged offence.
- The defence enters a plea for mitigation.
- Magistrates impose sentence.

If he pleads **not guilty**:

- The prosecutor outlines his case.
- The prosecutor calls prosecution witnesses who give evidence under oath.
- The prosecutor questions them (examination-in-chief).
- The defending solicitor questions prosecution witnesses (cross-examination).
- The prosecutor sums up his case.
- The defending solicitor presents defence case.
- The defending solicitor calls defence witnesses who give evidence under oath.
- The defending solicitor questions witnesses (examination-in-chief).
- The prosecutor questions defence witnesses (cross-examination).
- The defending solicitor sums up the case for the defence.
- Magistrates retire to consider their verdict.
- Magistrates announce their verdict:
  - if not guilty, the case is dismissed
  - if guilty, a sentence is imposed.
- If a fine is imposed it must be paid by the accused – it cannot be paid by his insurers even though they may pay the cost of the defence.

Magistrates have restricted powers of sentence – in most cases a fine not exceeding £5000 but in other serious cases, such as breaches of Prohibition Notices, they have power to impose a fine not exceeding £20 000 plus a custodial sentence of up to three months in gaol.

If a fine is imposed, it is usual for the defence to ask for time to pay – otherwise the fine must be paid before they leave the court.

If the unsuccessful party is aggrieved by the verdict they can give notice of appeal, which if allowed, sets the sentence to one side until the appeal is heard.

A visit to a Magistrate's Court will give a good feel for the types of cases heard and for the procedures followed.

1.4.2
County Court

Since 2005 the administration of the magistrates, Crown, county and Supreme Courts have been linked together. When a claim is submitted to the county court, a judge will decide which 'track' it is to take – multi-track, fast track or small claims track. If the claim is for less than £5000, it is likely to be allocated to the small claims track where the hearing is relatively informal and in private. Multi-track and fast-track cases are likely to be heard in open court which the public may attend.

Once a claim for more than £5000 has been submitted and the defence entered, there is normally a great deal of pre-trial negotiation. This is carried out by the solicitors from each side and involves the gradual sifting and agreeing of evidence. In the case of a claim for damages the procedure is as follows:

- The plaintiff files a request to the court for damages and gives particulars of the claim.
  - The claim can be based on:
      negligence
      breach of statutory duty
      both.
- The court issues a summons on the defendant.
- Both sides are represented by solicitors or barristers.
- The defendant enters a pleading refuting the claim.
- Contact is made between the solicitors for both sides regarding common evidence, exchanging information and endeavouring to reach a settlement.
- If agreement is not reached there may be a pre-trial review (interlocutory hearing) to clarify the issues, agree evidence to be used and attempt to find a settlement.
- The case goes to trial and follows a similar court procedure to that used in the Magistrate's Court.
- If an appeal is made it is directed to the Divisional Court or to the Court of Appeal.

The payment of any damages agreed or awarded is made by the employer's EL insurer who also bears the costs of the case. However, any fines imposed must be paid by the guilty party.

These court procedures are likely to change as a result of the Woolf Report (Access to Justice (1996) – dealing with civil case procedures) and the Auld Report (Review of the Criminal Courts of England and Wales (2001) – dealing with criminal court procedures). The aim of these changes is to simplify court hierarchy, provide shorter routes to case

'You expect me to believe that?'

resolution and give the courts greater control on representation. Specifically the recommendations set out to:

In civil cases:
- encourage alternative non-court routes to settle litigation;
- make litigation more co-operative and less adversarial;
- use single expert witnesses agreed by both parties;
- claims less than £5000 dealt with by judges in Small Claims Court;
- claims between £5000 and £15 000 start in County Court;
- major and complex cases start in the High Court.

In criminal cases:
- unify the Magistrates and Crown Courts;
- extend Magistrates powers of sentence to 12 months custodial;
- introduce an Intermediate Division;
- introduce a system of sentence discount to encourage guilty pleas.

1.4.3 Employment Tribunal

This hears cases arising from breaches of employment law where an employee complains against a decision of the employer in situations covered by the legislation. It also hears appeals against Improvement and Prohibition Notices issued by an enforcing officer. The tribunal is a civil court because the matters it considers relate to a difference of opinion between the employer and the employee, or the employer and the individual enforcing officer, over the interpretation of the law and not to a breach of the law.

The procedure in taking a complaint to an Employment Tribunal is as follows:

- The appellant sends notice of appeal in writing to the secretary of the tribunal.
- An appeal must be made within 21 days if it relates to an Improvement or Prohibition Notice.

- The respondent is sent a copy of the complaint.
- A date is set for the hearing.
- The tribunal consists of a legally qualified chairman supported by two lay members, one nominated from the employer's organization and one from the employee's organization.
- Both parties can present their own cases or can be represented (representatives need not be legally qualified but can be a friend such as a trade union official).
- The proceedings are under oath but much less formal than a court.
- Each side presents its case.
- The tribunal can ask questions.
- The tribunal makes its decision known, which is binding on both parties.
- No costs are awarded.
- Any appeal is made to the Employment Appeal Tribunal.

All court cases are time consuming (and hence expensive) for the employer, especially so in an Industrial Tribunal case where the employer is so much more closely involved.

1.5 Law-making in the UK	The process for making laws in the UK varies slightly between the statutes and the subordinate legislation, i.e. Regulations and Orders. All are in written form and copies may be obtained from TSO.
1.5.1 Statutes	The process for making statute law is as follows:

- The originator receives Parliament's permission to submit a proposal. The originator could be one of the major political parties or an MP (Private Member's Bill).
- A Green Paper outlining the aim of the proposal is prepared for discussion.
- A White Paper giving the policy and intent of the proposal is prepared.
- A Bill, which contains the proposed wording of the new statute, is presented to Parliament for its consideration.
- Parliament:
  (1) gives it a first reading
  (2) gives it a second reading
  (3) submits it to Committee for detailed examination of its content and implications
  (4) gives it a third reading.
- The Bill goes to the 'other house' (House of Lords) where a similar procedure is followed.
- When both Houses have agreed the content, the Bill is given a final reading before being submitted to the Sovereign for the Royal Assent which, by tradition, is never refused.
- On receiving the Royal Assent it becomes an Act and is entered in the Book of Statutes and becomes part of the statute law of the country.

Within an Act may be terms that empower a Minister of the Crown to make subordinate laws to supplement the main Act.

A subordinate law may be:

(1)   An Order:

- Usually concerning an administrative matter such as the timing of the implementation of particular sections of the main Act.
- Prepared with minimal consultation.
- Approved by Parliament by negative vote, i.e. no objections.

(2)   A Regulation:

- Prepared by the Health and Safety Commission/Executive (HSC/E) on behalf of the minister.
- Statutory requirement for interested bodies to be consulted.
- A Green Paper is prepared for limited discussion.
- A White Paper is prepared for limited discussion with interested bodies such as the CBI, TUC and affected industrial sector representatives.
- A Consultative Document is issued setting out the proposed wording, giving an explanation of the reasoning behind the proposal and calling for comments.
- Once comments are received, a draft regulation is prepared which may be subject to limited consultation with CBI, TUC and particular industry.
- A final proposal is submitted to the HSC for transmission to the minister.
- The minister lays the proposal before Parliament.
- It is approved by Parliament using either:
  - **negative voting procedure**, i.e. it lies on the table in the House for 30 days and if there are no objections it is automatically passed
  - **positive voting procedure**, i.e. it lies on the table of the House for 30 days during which time any MP can call for a debate on it. At the end of 30 days a positive vote is necessary to pass it.
- It then becomes a *statutory instrument* with the full power of law.

Acts and Regulations normally have a run-in period before full implementation. This allows TSO time to print copies and industry to adapt to the new requirements. In some cases, implementation may be phased over a number of months or years.

1.6 Law-making in the EU

In 1972 when the UK joined the then European Economic Community (EEC) now the European Union (EU), the UK government agreed to be bound by the various laws adopted by the Council of the European Communities. EU laws are 'adopted' by Council, not 'passed', and once adopted, they have immediate applicability on the government and its employees but do not apply to the private sector until national laws, incorporating the content of the directive, have been passed by Parliament.

Originally the Treaty of Rome required unanimous assent for a matter to be adopted, but the subsequent Treaties of Maastricht and Amsterdam have considerably modified the procedure. These Treaties introduced a *co-decision* procedure, shown diagrammatically in Figure 1.3, whereby the Council of the European Union (Council) and the European Parliament (EP) work on an equal footing and have to agree the text of proposed new EU legislation before it can be adopted. Where EP and the Council cannot agree, a Conciliation Committee has to be established to resolve any differences.

Under the social policy of the 1997 Amsterdam Treaty major areas to be tackled include:

- health and safety of workers
- working conditions
- rights to information and consultation
- equality between men and women

EU legislation comprises:

- Regulations
  - have direct applicability in the Member States at which they are aimed
  - apply mainly to the iron and coal industries and are rarely used for health and safety matters.
- Directives
  - the main body of EU law, used extensively in all areas of Community activities.
- Decisions
  - usually about a specific subject and may be aimed at a particular Member State where they have direct applicability.

In drafting health and safety directives there is wide consultation with interested sectors, both public and private. The European Parliament has a considerable say.

The drafting procedure follows the 'new approach to legislative harmonisation' whereby the directive itself contains broad objectives to be met and gives in annexes details of particular areas requiring attention, relying on *harmonized standards* (EN standards) to provide detailed requirements for conformity with the directive.

Figure 1.3 shows diagrammatically the process for promulgating a directive.

**1.6.1
European
harmonized
standards**

European standard-making bodies:

- Comité Européen de Normalisation (CEN), for mechanical standards.
- Comité Européen de Normalisation Electrotechnique (CENELEC), for electrical standards.
- Common Central Secretariat.

Their work is overseen by General Assembly and Technical Boards of representatives of the standard-making bodies of Member States.

EU commission programme of work

Proposal drafted by Directorate General (DG)
• DG III for equipment
• DG V for employment matters including health and safety

Draft referred to the Advisory Committee on Safety, Hygiene and Health Protection at Work for its opinion
• tripartite with representatives of government, employers and employees from each member state

DG submits proposals to the Commission

Commission publishes proposals in Office Journal 'C' series

Commission sends copies to:
• Council, who refer it to the Social Questions Working Group (SQWG)
• European Parliament (EP), whose response is drafted by the appropriate committee
• the Economic and Social Committee (EcoSoC), which is non-governmental and comprises representatives of employers, employees and other interested bodies

Council

| reject EP opinion and adopts a Common Position (CP) | accepts EP opinion and adopts directive |

Within 3 months EP may

| reject CP by absolute majority | amend CP by absolute majority | take no decision | approve the CP |

Council has 3 months to:

| reject EP amendments | approve EP amendments adopt directive |

Proposal fails

Council adopts directive

Referred to Conciliation Committee which has 6 weeks to obtain agreement

| fails to reach agreement | reach agreement on joint text |

Proposal fails

EP approves joint text by simple majority

Council adopts directive by QMV*

Figure 1.3 Diagram of the co-decision procedure for adoption of a directive

* A system whereby each Member State is given a number of votes depending on its size. 70% of votes in favour are required for adoption.

Member States include EU members, European Free Trade Association (EFTA) members plus affiliates from Eastern European and Middle Eastern countries.

The leading UK body is the British Standards Institution (BSI).

CEN and CENELEC are sponsored by the EU. They work closely with their international opposite numbers, the International Standards Organisation (ISO) and the International Electrotechnical Commission (IEC), respectively.

Harmonized standards take precedence over national standards and are recognized in the UK by a prefix to the standard number, i.e. BS EN.

Equipment conforming to harmonized standards and complying with the UK regulation that incorporates the requirements of the relevant EU directive, such as the *Supply of Machinery (Safety) Regulations 1998* and its amendments, can be marketed throughout the EU without restriction.

Harmonized standards are divided into four categories:

1 Type A, relating to fundamental safety concepts and principles

  – general safety principles
  – instruction handbooks
  – rules for drafting standards, etc.

2 Type B1 for safety **aspects** applicable to a large number of machines

  – safety distances
  – hand and arm speeds
  – noise and vibrations
  – hydraulic/pneumatic control systems
  – safety symbols, etc.

3 Type B2 for safety related **devices** that may be used on a variety of machines

  – two-hand controls
  – electro-sensitive safety systems
  – pressure-sensitive mats
  – interlocking devices, etc.

4 Type C for devices that are specific to certain types of machinery

  – cold forming of metal
  – industrial robots
  – mechanical handling equipment
  – construction equipment, etc.

2

Health and safety laws

The early health and safety laws, the *Factories Act 1961* (FA) and the *Offices, Shops and Railway Premises Act 1963* (OSRP), were aimed at correcting identified wrongs and tended to be very prescriptive in their content (i.e. they laid down how the wrong was to be corrected) and restrictive in their field by applying only to the particular operation, process or premises defined in the Act. Very few of the regulations made under these Acts are still in effect.

Health and safety laws have, in the past, evolved to protect against the hazards brought about by developing technology. While this still applies today, a greater emphasis is now put on the part that the employer can play. Attitudes have changed from protecting machinery to protecting people and are progressively looking at anticipating hazards (risk assessments) rather than waiting for them to manifest themselves through accidents.

Current health and safety legislation revolves round the *Health and Safety at Work, etc. Act 1974* (HSWA), which is proscriptive (i.e. it sets out the objectives to be achieved without specifying how). Regulations made under HSWA apply to all employments unless the regulation itself restricts its application.

HSWA places basic obligations on the employer, the employee, tenants in multi-occupancy premises and on the landlord. HSWA contains conditions that allow the making by the Minister of subordinate laws – statutory instruments known more commonly as Regulations and Orders – to enable particular requirements to be placed on a whole gambit of employment and workplace situations. These subordinate laws are building up to provide a great body of health and safety legislation that is superseding and updating the older well-established but outdated laws, and is more relevant to modern-day employment and processes.

The majority of health and safety regulations concerning employment at work, including those that emanate from the EU, are made by the appropriate government department (currently (2007) the Department of Work and Pensions for employment matters and the Department for Business Enterprise and Regulatory Reform for machinery). Drafting of the wording of the health and safety regulations is carried out by the Health and Safety Commission/Executive (HSC/E) under powers contained in HSWA.

To stimulate the growth of free trade in the EU a number of directives have been adopted that lay down safety requirements for goods and equipment. The drawing up of regulations to incorporate the content of these particular directives into UK laws is carried out by the Department of Business Enterprise and Regulatory Reform (DBERR) under powers contained in the *European Communities Act 1972*.

**2.1
A historical
perspective on
health and safety
laws**

This section covers very briefly some of the salient stages in the development of health and safety legislation in the UK.

1556	Part of a book on metal mining by Dr Agricola dealt with diseases of miners
1567	Treatise on diseases in mining and smelting by Dr Paracelsus
1690	*Boson* v. *Sandford* – first judgement that established the doctrine of vicarious liability of the employer, i.e. the employer being responsible for what his employees do while at work
1700	Book on trade diseases published by Italian physician Bernadino Ramazzini
late 1700s	Industrial Revolution
1784	Fever epidemic in Lancashire claimed many lives, particularly of young children
1795	Manchester Board of Health set up
1800	Combination Acts outlawing trade unions
1802	Act for the preservation of the health and morals of apprentices and others in cotton mills (the first health and safety Act)
1819	Further Act, prohibiting in mills the employment of children under nine years of age, reducing the hours of work for 9–16 year olds to twelve per day and prohibiting night work for under 16s
1824	Repeal of the Combination Acts
1832	Leeds doctor Charles Turner Thackrah published first English book on occupational diseases
1833	Act further reducing the hours of work of women and young persons, also authorizing the appointment of four Inspectors of Factories
1840 et seq	Saw the development of the doctrine of Breach of Statutory Duty as grounds for a claim in common law

Early industrial injury

1842	First Act relating to coal mines, banning the underground employment of women and children
1844	Further Act dealing with hours of work; also contained first requirement regarding machinery safety by prohibiting the cleaning of moving machines – but still applicable only to mills
1864	Safety legislation extended beyond mills to include some manufacturing factories
1871–5	850 boiler explosions resulting in death or serious injury
1878	Factory and Workshop Act – extending legal protection to nearly all manufacturing industries; first Act to consolidate safety legislation
1881	Boiler Explosions Act – laying down safety measures to be incorporated into boilers
1897	Workmen's Compensation Act – applicable to limited range of industries only
1901	Factory and Workshop Act – a further consolidating Act including, for the first time, powers for the Minister to make Regulations
1906	Workmen's Compensation Act – extended to all wage earners within certain wage limits
1937	A further consolidating Factories Act
1961	The last consolidating Factories Act
1963	The Offices, Shops and Railway Premises Act
1972	UK became a Member State of the European Economic Community and subject to EEC directives
1974	The Health and Safety at Work, etc. Act
1986	The Single European Act
1992	A series of Regulations incorporating the requirements of EEC directives into UK law, repealing considerable portions of the Factories Act 1961, the Offices, Shops and Railway Premises Act 1963 and revoking some associated Regulations
31st December 1992	The Single European Market established

Since 1992 the promulgation of health and safety laws has continued at an accelerating rate, many of them as a result of our membership of the EU.

2.2
The structure of health and safety laws

Legislation is structured in the following way:

Primary legislation

- Acts
  - known as statutes
  - fully debated in both Houses of Parliament
  - become law when they receive the Royal Assent.

Secondary legislation (sometimes referred to as subordinate legislation)

- Regulations
  - known as statutory instruments
  - drafting power delegated to a Minister of the Crown
  - interested bodies must be consulted
  - voted on by Parliament which gives them the power of law.
- Orders
  - known as statutory instruments
  - prepared by the Minister concerned and issued by him
  - deal with administrative matters such as implementation of an Act.
- Bylaws
  - made by local authorities under powers given to them by the Act which established them
  - they apply only within the local authority area
  - are not part of criminal law
  - can be challenged in court.

The following supporting documents can be used in the legislative process:

- Approved Codes of Practice
  - have a quasi-legal status
  - are drawn up by the HSC/E and approved by the relevant Secretary of State
  - lay down acceptable practices to give compliance
  - are official documents and are legally recognized in court without the need for *proving*
  - the courts may rely on them as evidence of a suitable standard to give compliance with statutory requirements
  - may be used in prosecutions as evidence of suitable means to give compliance
  - not exclusive or exhaustive but if other methods of compliance are used, the user may be called upon to justify the alternative method and to prove it gives compliance
  - can be used in court as evidence of acceptable levels of compliance
- various health and safety guidance booklets published by the Health and Safety Commission and Executive
- British and European harmonized standards
- industry-based standards often with HSE input.

Working to HSE guidance, British, European, international or industry based standards is usually deemed to give compliance.

Prior to 1974 the main health and safety Acts were the *Factories Act 1961* and the *Offices, Shops and Railway Premises Act 1963*. Both these Acts applied only in the premises defined in them. Similarly, any subordinate legislation made under either of them applied only in the premises covered by the main Act. However, legislation made since the coming into effect of the *Health and Safety at Work, etc. Act 1974* now covers all employment, both paid and voluntary, except domestic service.

Currently the content of EU Directives is incorporated into UK law by means of Regulations. Those concerned with employment, i.e. the use of machinery and equipment, are drafted by the HSC and made under HSWA. Those concerned with the standard of safety of goods are made by the Department of Business Enterprise and Regulatory Reform under the *European Communities Act 1972*.

2.3 The Health and Safety at Work, etc. Act 1974

The Health and Safety at Work, etc. Act 1974 (HSWA) is based on the recommendations of the Robens Report and sets out in principle the objectives to be achieved for ensuring high standards of health and safety in the workplace. This Act has superseded the earlier *Factories Act* and the *Offices, Shops and Railway Premises Act* and is underpinned by regulations made under its delegating powers.

A great deal of emphasis is placed on the role and responsibilities of the employer as the person who decides and controls what goes on in the workplace, but it does not forget that employees also have a contribution to make.

The general requirements for achieving high standards of health and safety at work are contained in ss.1–9 of HSWA, laying down broad objectives to be met and placing duties for meeting those objectives on the employer, the landlord, tenants, employees and suppliers.

The Act itself is divided into a number of parts:

Part 1

- Starts with a statement of the intent of the Act.
- Outlines the general duties placed on employers, landlords, tenants, employees and suppliers.
- Authorizes the setting up of the Health and Safety Commission and Executive and lays down how they should operate.
- Outlines the procedures to be followed in making regulations and issuing approved codes of practice.
- Lays down the enforcement powers of inspectors.
- Sets the criteria for obtaining and disclosing information that inspectors may acquire in the course of their investigations.
- Lists the offences under this Act and types of legal proceedings.
- Outlines appeals against action taken by an inspector.
- Denies the right of civil action based on breach of statutory duties placed by this Act.

Part 2

- Allows for the continuation of the Employment Medical Advisory Service.

Part 3

- Once dealt with building standards but has been repealed, its subject matter being incorporated into the *Building Act 1984*.

Part 4

- Deals with various administrative matters and interpretations.

Schedules

- Lists the circumstances and subject matters on which Regulations can be made (the list is pretty comprehensive and covers virtually all work situations).

More details of particular duties placed by this Act are dealt with in Part 2 of this book.

2.4 Health and safety law enforcement

The general responsibility for enforcing health and safety laws lies with the Health and Safety Commission (HSC) but the actual enforcement is carried out by the Health and Safety Executive (HSE). Currently (2007), the HSC and HSE are, effectively, separate entities but there are moves afoot to improve their effectiveness by combining them into a single organization. Because of the very wide range of areas to be covered, the HSE has delegated certain aspects or areas of enforcement to other bodies.

2.4.1 Enforcement responsibilities

- HSE – factories, manufacturing premises, fire prevention (where the high fire risk is in the process), construction sites, offshore industries, nuclear, agriculture, mines, quarries and Crown premises
- Local authorities – offices, shops, warehouses and non-machinery premises
- Fire Authority – the adequacy of fire risk assessments and fire safety standards (see s.18). However where the fire risk is in the process, enforcement is by the HSE.

2.4.2 Authority of inspectors

The work of enforcement is by inspectors:

- whose powers derive from ss.19 and 20 of HSWA
- who get their authority from a warrant they must carry when inspecting
- the warrant must be signed by a senior officer of the authorizing body.

2.4.3
Powers of
inspectors

HSE inspectors and local authority inspectors (Environmental Health Officers [EHO]) have the following general powers:

- Entry to any premises at any time work is being carried out.
- To obtain information:
  - take measurements and photographs
  - inspect documents but not 'privileged' documents
  - require persons to give information relevant to their investigations (failure to do so is an offence).
- Issue Improvement Notices.
- Issue Prohibition Notices.
- Not to pass to a third party information from investigations (to do so is an indictable criminal offence).
- Prosecute in a Magistrate's Court for alleged breaches of health and safety laws.

In exercising these powers they do the following:

- Carry out general inspections of workplaces (usually the more hazardous the workplace the more frequent the inspections).
- Investigate accidents at work.
- Investigate complaints about working conditions (this is given a high priority in their work).
- Give advice on specific health and safety problems.
- Pass to representatives of the employees factual information on what they have found.

An inspection

Fire officers have broadly similar powers except that they do not have authority to prosecute and their inspections are restricted to fire risk assessments and fire safety standards.

**2.4.4
Organization for
enforcement**

Notes on HSE enforcement organization (shown diagrammatically below):

1 Within the Regions are points of co-ordination of safety standards for particular industry sectors known as Sector Groups.
2 To advise it on matters for improving health and safety within an industry, the HSC has established a series of Industry Advisory Committees (IAC). These IACs are tripartite with the Sector Groups providing the secretariat.
3 Local authority inspectors, the Environmental Health Officers, have similar powers to an HSE inspector but report to the local authority. They may seek advice from the HSE on particular matters.
4 Fire Prevention Officers report through the Fire Authority to the Department of Communities and Local Government (DCLG) but work closely with the HSE on industrial matters.

Figure 2.1 The central administrative organization of the Health and Safety Commission and Executive (HSC/E)

Figure 2.2 Diagram of HSE's Field Operations Organization

2.5
The Management of Health and Safety at Work Regulations 1999

These regulations apply to all employment and supplement the requirements of HSWA and other extant legislation. Where there is overlap, the more stringent requirements take precedence.

They recognize that the major responsibility for health and safety at work lies with the employer but that the employee has a contribution to make (r.12). They place particular responsibilities on those who provide *contract, short-term* or *temporary* workers (i.e. the employment agencies) even though the control of how the work is carried out is not in their hands.

The main contents of the Regulations are:

Employers are required to:

r.3
- carry out suitable and sufficient **risk assessments** of any work activities that may put anyone at risk
- take particular account of the inexperience and vulnerability of young persons

- repeat the assessments if the work changes
- record the assessments if they have 5 or more employees [See Section 3.4 on risk assessment]

r.4 • employ a systematic approach when implementing preventative and protective measures

r.5 • implement arrangements for the effective planning, organization, control, monitoring and review of the preventative and protective measures taken
- if 5 or more employees, record those arrangements.

r.6 Health surveillance is to be provided to employees:

- where an occupational disease or health risk is identified
- if an identifiable disease or health risk is likely under normal foreseeable work conditions
- be carried out by a suitably qualified person, i.e. occupational health nurse or doctor.

r.7 Every employer must have:

- access to competent health and safety advice
- this service can be by a suitably qualified employee or as a bought-in service
- such advisers must be given time to fulfil their health and safety function
- where non-employees (consultants) are used:
  - they must be given information on any known or suspect hazards in the workplace and any special conditions that apply to the work
  - the qualifications required of such consultants are not specified but employers must satisfy themselves as to the consultant's competence.

(The qualifications required will vary with the size and complexity of the organization and its products. For an SME, the NEBOSH General Certificate may be adequate while for a major industry a Registered Safety Practitioner (RSP) or full corporate membership of IOSH may be required.)

r.8 Where danger exists in the workplace:

- those exposed must be told about it
- they must be told about the arrangements in place for their protection and how to avoid the danger
- employees must be trained in what to do in the event of imminent danger
- employees should be restrained from returning to their workplace while the danger still exists

* allowing for those with English language difficulties.

- there must be suitably trained staff to ensure the emergency procedures are followed – such as a fire marshal or a 'responsible person' appointed to deal with injury emergencies.

r.9 Employers should establish contact with the emergency services and inform their employees of the procedures for contacting them.

r.10 All employees are to be provided with 'comprehensible* and relevant' information on:

- the risks they face in their work including any that result from multi-occupancy of the premises
- precautions and preventative measures in place to protect them
- emergency procedures and who the emergency marshals are
- where a child is employed the parents must be given the above information.

r.11 Where two or more employers share the same premises (multi-occupancy):

- they must co-operate in meeting statutory obligations
- they must keep each other informed of specific risks arising from their particular operations
- these obligations may be met by appointing a safety co-ordinator for the premises
- where self-employed persons share premises they too must co-operate for the common good
- where someone other than the employer has control over part of the work premises (i.e. a landlord), he too must co-operate with his tenant employers to ensure health and safety in the various workplaces. This is in addition to any responsibility he carries for common areas under his direct control, i.e. entrance halls, stairs, lifts, etc.

r.12 Where non-employees carry out work whether as contractors, servicemen, or 'temps' from an agency, the client or person who has control of the area where they are to work must inform their employers, or the agency who sent them:

- of the work that is to be carried out
- of any risks associated with that work
- of the precautions and procedures in place to protect them
- and must ensure that the non-employees receive the same information.

r.13 In allocating work the employer must:

- take account of the employee's abilities, both mental and physical, to do the work required of him

---

* allowing for those with English language difficulties.

- ensure employees are adequately trained:
  - on joining the company
  - on exposure to new risks because of:
    * moving to a different job
    * the introduction of new techniques or materials
    * changes to existing equipment or the introduction of new equipment
    * the introduction of new systems of work
- ensure that the training:
  - is on-going
  - is repeated periodically (refresher training)
  - takes account of changing risks
  - takes place in working hours (or be paid for as overtime).

r.14 Employees must:

- follow the training they have been given when using any equipment, substance or safety device provided by the employer
- follow the employer's instructions to ensure legal requirements are met (see also s.7 HSWA)
- report to their employer any serious risks to health or safety they may find
- report any shortcomings in the safety arrangements.

r.15 Where temporary workers are employed, i.e. those on fixed-term contracts, they must be provided with information on:

- any special qualifications or skills necessary to do the work safely
- any health surveillance associated with the work.

Where temps from an agency are used, the agency must be provided with information on:

- any special qualifications or skills necessary to do the job safely
- any features of the job likely to affect the health and safety of the temporary workers
- The agency must pass this information on to the workers concerned and the employer must check before they start work that the agency has indeed passed the information on.

r.16 Where women who are of child-bearing age, pregnant or nursing mothers, the employer must assess the additional risks they face due to their condition and ensure they are:

- not employed in work that can put their child's health at risk
- move them to safer work
- reduce their hours of work or
- suspend them from employment.

r.17 Where a new or expectant mother works nights and a doctor or midwife certifies she should not follow that work, she should be suspended for as long as is necessary for her health and safety.

r.18 The employer need not take action if:

- the employee has not informed him in writing that she is pregnant, has given birth within the previous 6 months or is breast feeding
- the employee does not produce a certificate from a doctor or midwife confirming her condition
- the employee is no longer a new or expectant mother
- he cannot establish whether she remains a new or expectant mother.

r.19 Where young persons are employed, the employer must:

- ensure the young person is not put at risk because of:
  - inexperience
  - lack of awareness of dangers
  - physical immaturity
- not employ them:
  - where the work is beyond their physical and psychological ability
  - on processes likely to affect their health
  - where subjected to radiations
  - where experience is necessary to recognize danger
  - in extremes of heat or cold
  - in noisy areas
  - where exposed to vibrations
- provide competent supervision when training of young persons entails any of the above risks.

r.20 Empowers the Secretary of State to issue Exemption Certificates.

r.21 Maintains the employer's responsibility for the acts and defaults of his employees.

While these Regulations add considerably to the obligations placed on employers by HSWA and other Regulations made under it, they require no more than a caring employer would provide.

2.6 Other current health and safety legislation

A number of Acts not having direct relevance to health and safety, and not referred to in the above or the following text, nevertheless contain features that affect health and safety issues:

- *Environmental Protection Act 1990* deals with discharges to the atmosphere, to water courses and the disposal of waste on land. Some of the substances involved can pose a risk to the health of the community and to the ecology of the area.

- Various of the complex laws concerning employment which give protection to employees are relevant. In particular, s.100 of the *Employment Rights Act 1996* which gives employees protection against dismissal where health and safety is an issue. This Act has been amended by the *Employment Relations Act 1999* and the *Employment Act 2002.*
- *Sex Discrimination Act 1975* requires that women be given equal employment opportunities to men and that they are not discriminated against because of their sex except where to do so would cause a breach of another statutory requirement, such as handling certain chemicals that could put a woman's childbearing ability at risk. This Act has been amended by the *Employment (Sex Discrimination) Regulations 2005* and the *Equality Act 2006.*
- *Race Relations Act 1976* requires that there be equal treatment of all people regardless of race or ethnic origin. However, again exceptions can be made where a breach of other legislation may be caused, such as Sikhs not wearing hard hats on a construction site or having beards in certain food processes.

Further Acts that have health and safety connotations include:

- *The Consumer Protection Act 1987*
- *The Employer's Liability (Compulsory Insurance) Act 1969*
- *The National Insurance (Industrial Injuries) Act 1946*
- *The Occupier's Liability Act 1984*
- *The Offshore Safety Act 1992*
- *The Disability Discrimination Act 1995 as amended*

together with a developing range of associated Regulations.

Part 2 Management

The management of an organization is a complex mix of people and systems covering a very wide range of activities and functions. The function of management is to draw all these aspects together into a coherent whole and guide it towards the organization's goal. In health and safety, while personalities inevitably play a part, there are a number of established and documented facets of the part that management (or more truly the manager) can play in ensuring employees go home in the same healthy state in which they arrived for work.

This part looks at some of the management techniques that are important in achieving high standards of health and safety in the workplace.

3

Management responsibilities

While many managers may be remote from the work areas, they exert a great influence, both formally and informally, through 'setting the safety tone' for the organization. Attitudes in the boardroom are manifest on the shopfloor even though there is no direct communication link. Thus if the Board is concerned about health and safety this will show in a high level of safety performance in the organization as a whole.

Thus, in fulfilling their role in the enterprise, managers are responsible for ensuring that the part of the organization within their control is operating at maximum efficiency, not only in production and quality but also in health and safety.

This section looks at the health and safety responsibilities placed on managers and some of the techniques they can use to meet those responsibilities.

3.1 The role of management

Managers have a direct bearing on health and safety since they have control and can give instructions. Managers are also the focal point of a lot of employee attention and the manner of their behaviour (their example) and their 'seen' concern for health and safety matters is an important factor in employee attitudes.

Managers can influence safety performance by:

- setting policies that require high safety performance
- providing resources to achieve the aims of those policies
- ensuring that the resources provided are used properly and effectively
- giving local managers sufficient freedom and authority to achieve high standards of health and safety in their own way (encourage their initiative and commitment)
- holding local managers accountable for their safety performance
- demonstrating a commitment to safety by:
  - personal involvement in health and safety matters
  - encouraging high standards of safety by a proactive approach
  - ensuring health and safety matters are included on board agendas
  - giving health and safety equal consideration with production, finance and sales, etc.
  - being knowledgeable in health and safety issues when visiting the working areas and discussing them with employees.

In health and safety matters, the role of the employer and manager is to lead and control. 'How' is dealt with in the following sections.

3.2
Responsibilities for health and safety

UK laws put obligations on employers (i.e. the managing director, chief executive, partner or proprietor) to ensure the health and safety of employees. In large organizations the day-to-day performance of these obligations is delegated to subordinate managers who are given suitable authority to meet them.

Note that health and safety *responsibilities* of the employer cannot be delegated, it is only the *performance* or carrying out of those responsibilities that can be delegated.

The obligations placed on the employer by:

- the *Health and Safety at Work, etc. Act 1974* (HSWA)
- the *Management of Health and Safety at Work Regulations 1999* (MHSW Regs)
- the *Workplace (Health, Safety and Welfare) Regulations 1992* (WHSW Regs)

are qualified by the phrase 'so far as is reasonably practicable'. The meaning of this phrase was decided in 1949 by Lord Justice Asquith in *Edwards* v. *National Coal Board* [1949] 1 All ER 743 as:

> *'Reasonably practicable' is a narrower term than 'physically possible' and implies that a computation must be made in which the* quantum of risk *is placed in one scale and the* **sacrifice**, *whether in time, money or trouble, involved in the measures necessary to avert the risk is placed in the other; and that, if it be shown there is a gross disproportion between them, the risk being insignificant in relation to the sacrifice, the person upon whom the*

duty is laid discharges the burden of proving that compliance was not reasonably practicable. This computation falls to be made at a point in time anterior to the happening of the incident complained of.

Broadly interpreted this means that protective measures should be considered before any accident happens, and that if the cost of those protective measures is excessive for the benefits derived, the provision of the measures is not reasonably practicable. However, if this argument is used as a reason for not providing protective measures it may be necessary to justify it in court should an accident occur or should an inspector query it.

The obligations placed on employers by HSWA are *so far as is reasonably practicable* to:

- Make arrangements to ensure the health and safety of employees.
- Provide plant and equipment that is safe.
- Implement systems of work that are safe.
- Ensure the safe use, handling, storage and transport of both articles (equipment) and substances (chemicals).
- Keep employees and others (contractors, visitors, etc.) on the site informed on health and safety matters and arrangements.
- Provide adequate health and safety instruction and training.
- Ensure supervision is adequate and competent.
- Keep the workplace in good condition.
- Ensure the work environment does not put anyone's health at risk.
- Provide suitable welfare facilities.
- Have a written safety policy if five or more employees.
- If unionized, to recognize union-appointed safety representatives.
- Consult with safety representatives and employees on health and safety matters.
- Establish a safety committee when requested by two or more safety representatives (but there is nothing to prevent a voluntary safety committee being set up).
- In shared premises, to co-operate with neighbours on health and safety matters.
- Not emit noxious or offensive fumes or dusts.
- Not charge for PPE (personal protective equipment).

Duties are also placed on:

- employees:
  - to take care of themselves and others who may be affected by their acts or omissions
  - to co-operate with the employer in complying with statutory requirements
- everyone:
  - not to interfere with or misuse anything provided to meet a statutory requirement
  - report any dangerous situation to the local manager

- landlords:
  - to maintain common areas of premises under their control in a safe condition, i.e. entrances, gangways, stairs, lifts, etc.
  - to ensure any plant provided is safe to use
- suppliers of both substances and equipment:
  - carry out suitable tests to ensure equipment is safe for use
  - carry out tests to determine the chemical characteristic of a substance
  - provide the user with:
    * information on the limitation of its design
    * written instruction for its safe use
    * details of hazards of substances and precautions to be taken (safety data sheets).

Further obligations are placed on employers by MHSW and WHSW Regulations to:

- carry out risk assessments
- integrate health and safety into management systems
- provide health surveillance where the risk warrants it
- have available qualified health and safety adviser(s)
- take special precautions where pregnant women, nursing mothers and young persons are employed.

Where other more specific requirements arise those obligations are dealt with in the relevant part of the following text.

3.3 Safety policy

A duty is placed on employers (those with more than five employees) by s.2(3) of the HSWA to:

> ... *prepare and as often as may be appropriate revise a written statement of his general policy with respect to the health and safety at work of his employees and the organisation and arrangements for the time being in force for carrying out that policy* ...

Thus an employer must have in writing not only the policy statement but also details of the organization he has established for implementing that policy plus information on the rules and procedures by which the aims of the policy are be achieved.

Hence, meeting these obligations can be considered in three parts:

1 The **safety policy** should:

- state the aims of the organization for ensuring the health and safety of those who work in it or may be affected by its activities, i.e. contractors, visitors, neighbours, members of the public, etc.

- refer to measures that exist to consult with employees on safety matters
- indicate sources of expert safety advice
- refer to the means for disseminating health and safety information
- mention the importance of the part that the employees can play in achieving safe working conditions
- be:
    - in writing
    - signed by the head of the organization
    - dated
    - brought to the notice of all employees
    - monitored
    - reviewed periodically
    - re-issued as necessary.

2   The **organization** for implementing the policy should include:

- the name of the director with overall responsibility for health and safety
- the names of other members of the organization with safety responsibilities
- the safety responsibilities held by each member
- the responsibilities of subordinate managers for preparing health and safety policies for their departments
- relationships with recognized or other unions
- routes for joint consultation on health and safety matters
- specialized responsibilities for safety advice, training, monitoring the policy, etc.

3   The **arrangements** for achieving the aims of the policy should include:

- a list of the agreed safety rules and procedures with a brief summary of each
- existing safe systems of work
- provisions for safe maintenance
- procedure for carrying out risk assessments
- accident reporting and investigating procedures
- controls for the safe use of chemicals
- measures for the safe introduction of new machinery and chemicals
- arrangements for dealing with emergencies including evacuation
- methods for the dissemination of information
- training facilities
- procedures for joint consultation including safety committee meetings
- issue and use of personal protective equipment
- steps taken to protect the environment
- welfare arrangements and facilities
- any other safety related matter particular to the organization.

Encouraging employee
co-operation

For the safety policy, with its supporting arrangements, to be effective there must be means in place for checking that the agreed procedures and methods are being followed, that they are effective, and for implementing any changes those checks may show to be necessary.

3.4 Risk assessments

The term 'risk assessment' comes from the insurance industry and was one stage in their process of determining and spreading the liabilities they carried. It has been adopted into health and safety and its meaning widened to cover a spectrum of activities from the initial identification of a hazard to the establishment of safe working conditions.

It has become a central feature of new health and safety legislation and, while it was required by some earlier legislation, virtually all new legislation that is concerned in any way with matters that affect the health, safety and welfare of people at work does, and future legislation is likely to include a requirement to carry out a risk assessment.

Effectively, a **risk assessment** is a means whereby an employer can manage properly the risks faced by his employees and ensure that their health and safety is not put at risk while at work.

The Management Regulations place specific responsibilities on employers to:

- Identify hazards that pose risks to the health and safety of employees.
- Carry out 'suitable and sufficient' risk assessments of hazards identified.
- Decide what is 'suitable and sufficient' in the light of their operating circumstances.
- Cover in the assessment:
  - all equipment, both existing and new
  - materials and substances
  - any operations that could give rise to injury or damage.

- Give priority to protecting whole work force rather than individuals.
- Consider any risks from their operations that may affect non-employees such as agency and contract workers, contractors, visitors and those with a right of entry such as postmen, utilities employees, delivery drivers, etc.
- Give particular attention to the risks faced by young employees.
- Appoint an assessor:
  - to carry out assessments
  - who has knowledge of:
    * the work processes
    * health and safety legislation
    * current health and safety standards for the industry.
- Give the assessor time to carry out the assessments during working hours. (The assessor could be a supervisor or chargehand who has had health and safety training.)
- If more than five employees, record[†] the results of risk assessment.

**3.4.1
Definitions**

Certain terms are used in risk assessments:

hazard – something with the potential to cause harm
probability – the likelihood that the hazard will cause damage or harm
risk – a compound of the probability and the severity of the resulting damage or harm
danger – the state of being at risk
extent of the risk – a measure of the number of people likely to be affected and the severity of damage or harm, i.e. the consequences.

**3.4.2
Objective**

The object of a risk assessment is to identify hazards so that action can be taken to eliminate, reduce or control them before accidents occur that cause injury or damage.

However, it is not intended to stop activities that are enjoyable but carry a degree of risk. The Compensation Act 2006 allows courts to recognize the positive (enjoyment) value of an activity which has risks when deciding a common law claim.

In addition, the HSC's publication, *Principles of sensible risk management,* defines sensible risk management as:

- ensuring that workers and the public are properly protected
- providing overall benefit to society by balancing benefits and risks, with a focus on reducing real risks
- enabling innovation and learning, not stifling them
- ensuring those who create risks manage them responsibly and understand that failure to manage real risks responsibly is likely to lead to robust action
- enabling individuals to understand that as well as the right to protection, they also have to exercise responsibility.

---

† The method of recording is not specified but could be by diary note, special form or on disc. It is prudent to record all items viewed – the safe and the hazardous – for future reference and as evidence of the extent of the assessment.

It is **not** intended to:

- create a totally risk-free society
- generate useless paperwork mountains
- scare people by exaggerating or publicizing trivial risks
- stop important recreational and learning activities for individuals where the risks are managed
- reduce protection of people from risks that cause real harm and suffering.

3.4.3
Strategy

To achieve the objective and for risk assessments to be effective and workable, they need to be approached systematically. The following steps outline a logical and systematic approach:

1 Define the task or process to be assessed.
2 Identify hazards.
3 Eliminate hazards or reduce to a minimum.
4 Evaluate the risks from the residual hazards.
5 Develop precautionary strategies.
6 Train operatives in new work methods.
7 Implement precautionary measures.
8 Monitor performance.
9 Review periodically and revise as necessary.

Considering each of these steps separately:

1 Prepare a programme of risk assessments.

- list all the tasks, processes and work areas that present hazards
- arrange the list in order of descending level of hazard and draw up a programme of risk assessments
- implement the programme employing the following procedures to reduce and evaluate each hazard.

2 Identify hazards.

Techniques:

- safety inspections (a general safety survey of the workplace)
- safety tours (identify hazards along a fixed pre-determined route)
- safety sampling (check for only one type of hazard, then later repeat for other hazards)
- safety audit (a count of the numbers of the different hazards found for comparison with similar earlier or later audits)
- environmental surveys
- accident reports
- reports of near misses or 'close shaves'
- comments from employees
- reports in the press and from trade associations.

3 Eliminate or reduce hazards.

Possible action:

- eliminate hazardous operation or material (the matter then ends since the hazard no longer exists)

- for those hazards that cannot be eliminated:
  - develop safer working methods
  - use alternative less hazardous materials.

4 Evaluate the residual risks, i.e. carry out a risk assessment.

Considerations:

- level or extent of hazard faced
- time of exposure
- number of people exposed
- probability of event happening.

Evaluate either qualitatively:

- based on personal estimation (e.g. high, medium or low)
- essentially subjective
- can be given numeric values (see below).

or quantitatively:

- based on published data of failure rates
- often quoted as probabilities.

5 Develop precautionary strategies.

Techniques:

- elimination of hazardous equipment, materials, substances or method of working
- substitute safer equipment, materials, substances or working methods
- prevent exposure or contact by use of bulk supplies, containment or the provision of suitable protective means (safeguards)
- control exposure or contact by restricting access to or time of contact with substance
- provide PPE as a last resort.

6 Train operatives:

- in new working methods
- in proper use of precautionary measures.

7 Implement precautionary measures.

8 Monitor performance.

Ensure the following:

- precautionary measures/working methods are being used
- the precautionary measures are effective
- the new work methods have not created new hazards
- possible weaknesses in the measures are highlighted and corrected.

9   Review and revise.

- to ensure methods still effective
- to bring precautions up-to-date
- whenever work methods or materials change
- when existing assessment becomes no longer effective.

Figure 3.1 illustrates this process diagrammatically.

Figure 3.1 Diagram of the process of risk assessment

3.4.4
Risk rating

Where a number of residual hazards remain there may be a need to determine priorities for dealing with them. This can be done either subjectively based on the assessor's knowledge of the processes or by evolving, for each residual hazard, a numeric value and hence a risk rating by considering a number of factors:

Factor	Extent	Value
Hazard	• unlikely to cause injury	1
	• may cause minor injury	2
	• could cause injury requiring first aid	3
	• could cause injury requiring medical treatment	4
	• could cause major injury	5
	• life threatening, possibility of fatality	6
Probability	• most unlikely	1
	• remote possibility	2
	• reasonably possible	3
	• fairly likely	4
	• very likely	5
	• almost certain	6
Severity	• negligible injury	1
	• minor injury	2
	• major injury	3
	• multiple injuries	4
	• single fatality	5
	• multiple fatalities	6

The risk rating is obtained by multiplying together the values given to each of the factors:

hazard value × probability value × severity value = risk rating

This gives a numeric value to each of the hazards which, as well as highlighting the highest risk, provides the basis for a priority rating.

3.4.5
Design risk assessment

The above deals with risk assessment in respect of existing plant and premises. However, risk assessments at the design stage are becoming increasingly demanded. These require specialist knowledge of and experience in the operation of the particular plant and are often carried out by multi-discipline teams.

Typical techniques include:

- 'WHAT-IF' technique of assessing the likely effects of foreseeable faults.
- Fault Tree Analysis (FTA), developing a logic diagram to trace back possible faults.
- Hazard Analysis (HAZAN), analysing the effects of possible faults.
- Hazard and Operability Studies (HAZOPS), involving a multi-discipline group considering the effects of faults.
- Failure Mode and Effect Analysis (FMEA), studying the effects of component failures.

A more detailed consideration of design risk assessments is given in 'Safety with machinery' (ISBN–10: 0–7506–6780–X).

3.4.6 Legislation

Existing legislation that has a risk-assessment requirement includes:

- *The Control of Asbestos at Work Regulations 2002*
- *The Control of Lead at Work Regulations 2002*
- *The Control of Substances Hazardous to Health Regulations 2002*
- *The Management of Health and Safety at Work Regulations 1999*
- *The Personal Protective Equipment at Work Regulations 2002*
- *The Manual Handling Operations Regulations 1992 (as amended)*
- *The Health and Safety (Display Screen Equipment) Regulations 1992*
- *The Dangerous Substances and Explosive Atmospheres Regulations 2002*
- *The Control of Noise at Work Regulations 2005*
- *The Work at Heights Regulations 2005*
- *The Regulatory Reform (Fire Safety) Order 2005*
- *The Provision and Use of Work Equipment Regulations 1998* in its requirement to determine safety of use 'so far as is reasonably practicable'
- *The Supply of Machinery (Safety) Regulations 1992* in the 'essential safety requirement' to *eliminate or reduce risks as far as possible.*

3.4.7 Sources of information

Major sources of information are the Approved Codes of Practice associated with the above Regulations. These are available from HSE Books:

L5	Control of substances hazardous to health. The Control of Substances Hazardous to Health Regulations 2002. Approved Code of Practice and guidance.
L21	Management of health and safety at work. Management of Health and Safety at Work Regulations 1999. Approved Code of Practice and guidance.
L22	Safe use of work equipment. Provision and Use of Work Equipment Regulations 1998. Approved Code of Practice and guidance.
L23	Manual handling. Manual Handling Operations Regulations 1992. Guidance on the Regulations.
L24	Workplace health, safety and welfare. Workplace (Health, Safety and Welfare) Regulations 1992. Approved Code of Practice and guidance.
L25	Personal protective equipment at work. Personal Protective Equipment at Work Regulations 1992. Guidance on the Regulations. (Due to be updated to cover PPE Regulations 2002)
L26	Display screen equipment at work. Health and Safety (Display Screen Equipment) Regulations 1992. Guidance on the Regulations.
L27	Work with asbestos which does not normally require a license. Control of Asbestos at Work Regulations 2002. Approved Code of Practice and guidance.

L132   Control of lead at work. The Control of Lead at Work
          Regulations 2002. Approved Code of Practice and guidance.
HSG 97   A step by step guide to COSHH assessment
IND 163   5 steps to risk assessment

**3.5
Techniques of
hazard
identification**

The identification of hazards before they cause an accident is central to all accident prevention activities. However, hazard identification is not an exact science but a subjective activity where the measure of the hazard identified will vary from person to person depending on their experiences, attitude to risks, familiarity with the process, etc. By repeating, or employing a range of identification techniques, the number of residual hazards will be reduced. It is doubtful if they will all ever be totally eliminated.

The findings of each inspection should be recorded so they can be referred to when deciding remedial action needed and for comparison with previous inspections.

There are a number of identification techniques from which to select the one that is likely to be most effective in a particular organization or which will provide the information required in respect of a particular process. They include:

1   Safety surveys

    • Sometimes called safety inspections.
    • Entail a general inspection of the whole work area.
    • Tend to be less detailed than other techniques.
    • Do give an overall picture of the state of accident prevention across the particular work area.

2   Safety tours

    • Inspection is restricted to a predetermined route.
    • Need to plan subsequent routes to ensure complete coverage of work area.
    • Reduces the time taken by each inspection.

3   Safety sampling

    • Looks at only one aspect of health or safety.
    • Concentrates the mind and identifies more detail.
    • Need to plan a series of samplings to cover all aspects of health and safety.

4   Safety audit

    • Detailed inspection of the workplace.
    • Seeks to identify all types of hazard.
    • Numbers of each type of hazard identified should be recorded.
    • Can be developed to give a numerical score to measure the 'safety health' of the organization.

- By repeating the audit a measure of improvement, or otherwise, in safety and health can be obtained.
- Can be time-consuming.

5 Environmental checks

- Based on measurements of concentrations of chemicals in the atmosphere.
- Can identify possible health hazards faced in workplace.
- Recording of sequential readings can show improvements or otherwise.
- Checks by 'grab sampling' are not very accurate and can be expensive.
- Electronic instruments expensive to buy but give instantaneous accurate reading.
- Electronic instruments can be used continuously over a long period.

6 Accident reports

- Post accident recording.
- Need to include minor as well as lost time injuries.
- Information obtained from accident report.
- Report should give indication of preventative action needed.

7 Near-miss reports

- Reports of incidents that in slightly different circumstances could have caused an accident.
- Needs the right safety culture to be effective.

8 Feedback from employees

- Can be formally through a safety committee or informally to supervisor.
- Needs a 'no-blame' culture to encourage employees to report matters.
- Employees often know and can say what needs to be done.
- Needs feedback to employees on action taken to retain management credibility.

3.5.1
Typical matters for inspections

- Welfare facilities:
  - canteen
  - toilets
  - first aid
  - smoking arrangements

- Fire precautions:
  - extinguishers
  - escape routes
  - alarms and fire drills

- Machinery:
  - guarding
  - following agreed system of work

      – state of machinery
      – reports of statutory examinations
      – compliance with legislative requirements

- Working conditions:
  - temperature
  - lighting
  - cleanliness and housekeeping
  - fumes and dusts
  - general decoration

- Access and gangways:
  - well marked
  - not encroached on
  - surface condition
  - adequately lit.

3.6 Organization for safety at work

The object of organization within an enterprise is to provide the means by which the aims of that enterprise can be achieved. As far as safety is concerned, it must make provision for, not only its own employees, but also for visitors to the premises, contractors working on the premises, invitees, neighbours and members of the public who may be affected by the way the enterprise operates.

3.6.1 Techniques

Techniques an organization can utilize to ensure its efforts are effective include:

- leading by example
- good communication
- effective joint consultation
- commitment from all
- generating identity with the organization
- involvement and participation
- work and job design
- competitive payment systems
- commitment to quality
- satisfying the customer.

3.6.2 Types of organization

Formal – the organization's structure is dictated by directors as being that which is necessary to meet commercial aims.

Informal – casual groups of individuals band together, linked by common interests. They decide amongst themselves how they will behave in the environment in which they work and fix their own work targets, often in conflict with the demands of the formal organization. Compliance with the norms set by these informal groups is often given priority over those imposed by the formal organization.

Official – i.e. governmental departments; in health and safety these are the Health and Safety Commission and Executive.

Professional – the Institution of Occupational Safety and Health, the Chartered Institute of Environmental Health Officers, British Occupational Hygiene Society, etc.

3.6.3
Roles in the organization

Within an organization the health and safety role played varies according to the person's position within the hierarchical structure. Thus:

- Managing director
  - sets the tone for the organization by his attitude, commitment and involvement
  - controls resources and ensures that facilities are available for achieving the safety policy aims
  - makes resources available for health and safety matters and ensures they are used effectively
  - leads by example
- Production manager
  - responsible for organizing the work and ensuring it is carried out safely
  - consults with the employees on health and safety matters
  - agrees safety rules and practices
  - ensures those safety rules are followed
  - provides the means (labour, materials and finance) to achieve and maintain a safe workplace
  - chairs the safety committee
  - leads by example
- Safety adviser
  - advises managers on all matters of health and safety
  - ensures that what needs to be done is done safely
  - does not interfere with production work except in an emergency
  - organizes safety committee meetings but does not take the minutes
  - acts as contact with outside safety organizations such as HSE, RoSPA, BSC, local safety groups
- Foreman/chargeman/supervisor
  - front-line manager
  - checks that safety rules are being followed
  - ensures machinery and equipment are safe to use
  - initiates discipline for breaches of safety rules
  - initiates requests for safety work
- Safety representative
  - represents employees on safety matters
  - carries out inspections and investigations with the agreement of the manager
  - ensures employees follow the safety rules
- Safety committee
  - considers reports from the manager and safety adviser on safety matters
  - comments on safety standards and practices and makes recommendations for improvements
  - keeps employees informed on safety matters

- The company safety organization and performance should reflect the organization and performance elsewhere in the company, especially where BS EN ISO 9001 and 14001 registrations exist.

3.6.4
Safety
organizations

- The Institution of Occupational Safety and Health (IOSH)
  - the recognized professional body
  - sets professional standards for practising safety advisers
  - represents the interests of the safety practitioner
- The National Examination Board in Occupational Safety and Health (NEBOSH)
  - the recognized examining body for occupational health and safety subjects
  - independent and self-financing
- National Training Organisation for Employment (Employment NTO)
  - responsible for the development of Vocational Standards of Competence (NVQs) in various areas of employment including health and safety
- The Royal Society for the Prevention of Accidents (RoSPA)
  - largest UK safety organization
  - covers safety in all work and leisure activities
  - provides training and safety advice
  - organizes major safety exhibitions
- British Safety Council (BSC)
  - independent safety organization
  - provides training and safety advice
  - has strong safety lobby
- British Standards Institution
  - not strictly a safety organization but many of its standards contain safety requirements
- Industry safety bodies
  - organized within and by particular industries
  - often in co-operation with the HSE through Industry Advisory Committees
  - set standards particular to their industries
  - voluntary and rely on employers to implement the standards.

3.7
Promoting health
and safety in the
workplace

There are a number of techniques that can be employed to improve and promote effective levels of health and safety in the workplace that complement legislative requirements and are good industrial and commercial practice. They aim to increase awareness of the need for high standards of health and safety at work.

Typical techniques include:

- Evaluation of safety knowledge
  - Complete Health and Safety Evaluation (CHASE)
  - International Safety Rating System (ISRS).
- Risk assessment
  - identify hazards and remove them or take appropriate precautions.

- Monitoring safety standards
  - safety inspections and surveys that are general in nature and cover whole workplace
  - safety tours that follow a predetermined route and note safety items
  - safety audits comprising detailed examination and quantification of safety items
  - safety sampling that looks only at one specific aspect of health or safety.
- Communicating the safety message by:
  - poster
  - news sheet
  - tool box talks
  - personal example.
- Using safer processes or materials.
- Including health and safety as inherent part of skill training.
- Keeping all plant well maintained
  - planned maintenance.
- Developing and using safe systems of work.
- Ensuring supervisors are trained and competent in health and safety matters.
- Practice of emergency drills and procedures.
- Providing good working conditions and environment.
- Developing and maintaining effective lines of joint consultation.

3.7.1 Safe systems of work

A **safe system of work** is a considered method of working that takes proper account of the potential hazards to employees, and others such as visitors and contractors, and provides a formal framework to ensure that all steps necessary for safe working have been anticipated and implemented.

All systems of work should be safe, but where the risk is such that unambiguous working instructions are necessary, the system of work should be in writing.

Typical procedure for developing a safe system of work:

1 Identify hazards from:
   - energy
     - electricity
     - steam or high pressure hot water
     - compressed air
     - hydraulic systems
     - compressed springs (in cylinders, robots, etc.)
   - materials
     - corrosives
     - asphyxiant gases
     - flammables and explosives
     - toxic substances
   - plant
     - machinery
     - cranes and lifting equipment
     - internal transport

- dangerous places
  - working at heights
  - in confined spaces
  - strange environment.

2 Remove danger
  - change process and/or materials.

3 Provide protection
  - guards
  - personal protective equipment.

4 Develop safe system of work
  - in writing
  - use:
    * locking off
    * permit-to-work.

5 Provide suitable training.

6 Provide special equipment
  - harnesses
  - breathing apparatus/masks
  - ear muffs/plugs
  - safe working platforms.

7 Monitor that the system is being followed.

A useful mnemonic device is 'IRPSTEM':

I – Identify hazards
R – Remove dangers
P – Provide protection
S – Safe system of work
T – Training
E – Equipment
M – Monitor.

**3.7.2
Permit-to-work
system**

Certain operations present higher than normal risks and require stricter controls. In these cases a 'permit-to-work' system should be implemented. A **permit-to-work** procedure is a formal written system of authorization used to control certain types of work which are potentially hazardous.

The permit-to-work must be in writing and each part authorized by a suitably responsible person. The essential parts of a permit-to-work are:

1 Statement (description) of the work to be undertaken and the safety precautions to be implemented before work can commence.
2 Confirmation that the precautions have been taken and it is safe to commence the work.
3 Statement that the work is complete and it is safe to return the plant to production.

X Y Z Company Limited
PERMIT-TO-WORK

NOTES:
1 Parts 1, 2 and 3 of this Permit to be completed before any work covered by this permit commences and the other parts are to be completed in sequence as the work progresses.
2 Each part must be signed by an Authorized Person who accepts responsibility for ensuring that the work can be carried out safely.
3 None of the work covered by this Permit may be undertaken until written authority that it is safe to do so has been issued.
4 The plant/equipment covered by this Permit may not be returned to production until the Cancellation section (part 5) has been signed authorizing its release.

PART 1 DESCRIPTION

(a) Equipment or plant involved _____

_____

(b) Location _____

(c) Details of work required _____

_____

Signed _____   Date _____
person requesting work

PART 2 SAFETY MEASURES

I hereby declare that the following steps have been taken to render the above equipment/plant safe to work on: _____

_____

Further, I recommend that as the work is carried out the following precautions are taken: _____

Signed _____   Date _____
being an authorized person

PART 3 RECEIPT

I hereby declare that I accept responsibility for carrying out the work on the equipment/plant described in this Permit-to-Work and will ensure that the operatives under my charge carry out only the work detailed.

Signed _____   Time _____   Date _____

Note: After signing it, this Permit-to-Work must be retained by the person in charge of the work until the work is either completed or suspended and the Clearance section (Part 4) signed.

PART 4 CLEARANCE

I hereby declare that the work for which this Permit was issued is now completed/suspended* and that all those under my charge have been withdrawn and warned that it is no longer safe to work on the equipment/plant and that all tools, gear, earthing connections are clear.

Signed _____   Time _____   Date _____
* delete word not applicable

PART 5 CANCELLATION

This Permit-to-Work is hereby cancelled

Signed _____   Time _____   Date _____
being a person authorized to cancel a Permit-to Work

Figure 3.2 A Permit-to-work

In carrying out this procedure it is essential that:

- there is clear understanding of who:
  - may authorize each part
  - is responsible for specifying the necessary precautions.
- proper training and adequate instructions are given in:
  - the work to be carried out
  - the procedure to be followed in the issue and use of the permits.
- the work is monitored to ensure the laid down procedures and methods are being followed.

The objects of the procedure are to:

- reduce to a minimum the risk of injury or ill-health to those carrying out the work
- ensure proper authorization for the procedures and work
- make clear to those carrying out the designated work:
  - the exact identity, nature and extent of the job
  - the hazards faced, the precautions to be taken
  - any limitations on the extent of the work or time allowed
- ensure that the person in charge of the plant or area is aware of all the work that is to be done
- provide a record of the nature of the work, the precautions taken and the people involved
- provide a formal hand-back procedure to ensure that the part of the plant affected by the work is in a safe condition to return to production
- A typical Permit-to-Work is shown in Figure 3.2.
- See also HSC booklet HSG 250 *Guidance on permit-to-work systems.*

4   Human resources

One part of the work environment that has, in the past, been taken very much for granted is the workforce, even though they are probably the most costly component in a company's budget. In recent years this attitude has changed and there is now a great deal of attention focused on and legislation aimed at protecting employees, not only from the dangers met at work but also in the security of their employment and in giving them a greater say in the running of the enterprise.

The following sections deal with those aspects of health and safety that are related specifically to employees in their various roles.

4.1 Health and safety training

Obligations are placed on employers by s.2(2)(c) of HSWA:

. . . to provide such information, instruction, training and supervision as is necessary to ensure, so far as is reasonably practicable, the health and safety at work of employees.

This is enlarged on in Regulation 13 of the *Management of Health and Safety at Work Regulations 1999* which requires employers to:

- take account of the ability of employees to perform their tasks
- provide training when:
  - employees:
    * first join the company
    * are transferred to a different job
    * are given changed responsibilities
  - working methods of existing equipment are changed
  - new equipment is introduced
  - new technology is introduced
  - new materials are used
  - the system of work is changed
- give revision training periodically
- give training during working hours.

Where contract staff are hired on a temporary basis and are put to tasks for which training is considered necessary for full time employees, the employer is required to give those temporary staff the same training.

Where contractors are brought on to the premises to carry out work, they too must be told of any particular risks they may face and trained in the techniques to avoid those risks.

4.1.1
Training content

Induction training

- For all employees:
  - to cover all aspects of employment including:
    * fire precautions and evacuation drill
    * the company's safety policy
    * details of company products
    * tour of the premises to identify location of facilities
    * occupational health facilities such as first aid arrangements
    * accident prevention activities
    * safety rules to be obeyed
    * fire and security arrangements
    * who to contact if in doubt.
- For contractor and his employees:
  - local safety rules to be obeyed
  - any special hazards in their area of work
  - safe systems of work and permit-to-work systems procedures
  - emergency and evacuation procedures
  - allowed access routes
  - permitted use of welfare facilities – canteen, toilets, first aid, etc.

On-going health and safety training

- For all employees:
  - re-affirmation of safety rules
  - hazards likely to be met in the work and techniques for avoiding them
  - safety devices and how to use them
  - issue, use and maintenance of PPE
  - re-affirmation of emergency and evacuation procedures
  - action in the event of an accident
  - procedure on identifying a hazard.
- For supervisors
  - more detailed instruction on legal requirements – HSWA, regulations and approved codes of practice
  - common law duties of care
  - techniques of risk assessment
  - getting benefit from safety inspections
  - accident investigation
  - fire precautions and special responsibilities
  - need for personal and special hygiene requirements
  - safety committee and the role of the safety representative
  - employee relations in health and safety.
- For managers
  - interpretation of health and safety legislation
  - economics of safety:
    * cost of accidents
    * employer's liability and accident claims

     * risk management
     * cost benefits of safety and provision of safeguards
    – safety culture
    – safety inspections and audits
    – monitoring safety performance
    – motivating for safety
    – integration of safety as inherent part of all work.
   • For safety representatives (whether union-appointed or voluntary)
    – their role in health and safety in the workplace
    – their rights regarding:
     * special training
     * time to carry out inspections
     * investigation of accidents
     * membership of safety committee.

**4.1.2
The training
process**

Before training is given, the need for it should be assessed and the programme of training tailored to meet the identified needs by:

- analysing the training needs
- drawing up a training plan
- developing a training programme
- setting training objectives (what the programme should achieve)
- briefing speakers/trainers
- preparing training aids and visual aids
- implementing the programme as a training course
- evaluating the effectiveness of the training (de-brief)
- revising the programme, speakers, training aids for the next course.

**4.1.3
Training techniques**

The particular technique used should match the level of course being given but could include:

- lectures and talks
- videos and films
- role-playing by course members
- case studies – report back to course
- syndicate discussions
- practical exercises either out on site or using a table-top model
- on-the-job.

Training is an investment for the future whether it be to improve skills, utilize the latest technology, give employees greater job satisfaction or to ensure they go home in one piece.

**4.2
Young persons
at work**

Children:

- under 13 years of age are prohibited from employment
- between the ages of 13 and 16 may not be employed in industrial processes unless on a training scheme approved by the local authority

- aged between minimum school leaving age (around 16th birthday) and 18 years, generally known as 'young persons', may be employed but under strictly controlled conditions.

Youngsters leaving school to start work are vulnerable because:

- they are entering a strange environment that is alien to what they have known so far
- they have no built-in recognition of the dangers of machinery or equipment
- they are at a stage of physical development that leaves them susceptible to a range of chemical substances and physical hazards that may cause lifelong disabilities or effects.

It's never too early to start safety training

Particular conditions concerning the employment of young persons are contained in MSHW.

r.1   Defines:

- *child* as a person who is not over the compulsory school age
- *young person* as any person who has not attained the age of 18.

r.3   Requires that a young person shall not be employed unless:

- the employer ensures his risk assessment allows for the young persons's vulnerability taking special account of:
  - inexperience, lack of awareness and immaturity
  - the suitability of work equipment and workplaces
  - possible exposure to hazardous substances
  - the organization of processes and activities
  - the extent of health and safety training provided.

r.10   Lays down that where a child is to be employed information must be provided to its parents on:

- the risks identified by the risk assessment
- the precautions and preventative measures in place
- any other risks arising from the way the work is organized.

r.19   Requires young persons to be protected because of:

- their lack of experience
- lack of awareness of existing or potential risks
- their lack of training

They must not be employed on work that:

- is beyond their physical and psychological capacity
- exposes them to substances that:
  - are carcinogenic
  - can cause genetic damage
  - can cause harm to an unborn child
  - can have a chronic effect on health
  - are radioactive
- exposes them to:
  - extreme heat or cold
  - noise
  - vibration
- makes them vulnerable due to their:
  - inexperience
  - lack of training
  - inability to recognize hazards and risks

Except where the young person is no longer a child and:

- it is a necessary part of training
- is competently supervised
- any risks have been reduced so far as is reasonable.

4.2.1
Problems faced
by young persons
starting work

In making the transition from school to work young persons need to adapt to:

- change from 5½-hour day to 7½- or 8-hour day
- monotony and possibly boredom until work understood
- lack of breaks every 40 minutes
- unfamiliar equipment and machinery
- unfamiliar hazards
- safety rules to follow
- safety devices to learn about
- protective clothing.

4.2.2
Training for young
persons

- Same induction training, as for all employees, plus emphasis on:
  - need for good housekeeping
  - no running
  - no horse play or practical jokes
  - no short cuts:
    * in the job
    * through the workplace
  - role of supervisor as trainer
  - if in doubt, ask
  - where to get information and advice
  - explain 'why fors'
  - follow the example of the supervisor
  - safe use of machinery and equipment
  - proper use of PPE.

4.2.3
Supervision

Both statute and common law place a special duty on employers to ensure young workers are properly and adequately trained and supervised. Supervision is crucial to ensuring the safety of youngsters, especially in their early days at work and when being put to work on machinery of any sort. The supervisor should remain in the vicinity of the youngster to keep an eye on what is being done and to be instantly available to give advice or to prevent bad habits from developing. The supervisor must be competent in the work being done.

4.2.4
Special risks

Where work involves any of the following processes or materials, special attention should be paid to training, protecting and supervising young persons:

- carriage of dangerous goods
- dangerous (heavy) metals
- dangerous machinery
- driving work vehicles, cranes, etc.
- explosives
- high fire-risk processes
- ionizing radiations
- lead

- manual handling
- power presses
- power-driven machinery
- toxic substances
- woodworking machines (effectively prohibited unless under supervised training).

In addition, legislation places severe restriction on young persons working in:

- agriculture
- potteries
- asbestos
- nuclear
- certain parts of the chemical industry.

Typical of the machines at which young persons should not work without proper training and supervision were listed in the *Dangerous Machines (Training of Young Persons) Order 1954*: While this Order has been revoked, it does provide a useful list of machines at which special precautions are necessary if young persons are employed.

1   Machines driven by power:

- brick and tile presses
- machines used for opening or teasing in upholstery or bedding works
- carding machines in use in the wool textile trades
- corner staying machines
- dough brakes
- dough mixers
- worm pressure extruding machines
- gill boxes in use in the wool textile trades
- the following machines in use in laundries:
  – hydro-extractors
  – calenders
  – washing machines
  – garment presses
- meat mincing machines
- milling machines in use in the metal trades
- pie and tart making machines
- power presses including hydraulic and pneumatic presses
- loose knife punching machines
- wire-stiching machines
- semi-automatic wood-turning lathes.

2   Machines whether driven by power or not:

- guillotine machines
- platen printing machines.

Particular care and attention needs to be paid in the training of young-sters starting work to correcting any bad working habits since they will be carried throughout their working life.

| 4.3
Joint consultation | In endeavours to ensure high levels of health and safety in the workplace, use must be made of all the know-how available. Where matters relating to the actual work areas are concerned, the people best qualified to comment are those who work there. In many cases they see hazards and problems and get round them long before they reach a manager's notice. Also where a problem arises concerning particular aspects of the workplace, the local workers are well placed and well qualified to offer helpful suggestions for its solution. This potential needs to be tapped. |

Joint consultation

In addition, if employees are consulted before changes are made they will accept them that much more readily.

Joint consultation can take many forms and routes depending on the culture and environment in the workplace. Rights to appoint safety representatives have been given to recognized trade unions who are under no pressure to exercise this right which remains should they feel the need to call on it later. These rights are embodied in the *Safety Representatives and Safety Committees Regulations 1977*.

Similar rights exist for volunteer or management appointed safety representatives under the *Health and Safety (Consultation with Employees) Regulations 1996* as amended. It has been found that with good industrial relations and an effective voluntary safety committee, the unions are happy not to exercise their rights of appointment.

**4.3.1
Safety
representatives**

The rights of safety representatives include:

r.3 The right to be consulted on:

- the introduction of measures affecting health and safety
- arrangements for appointing safety adviser
- arrangements for appointing fire and emergency wardens
- the provision of health and safety information relevant to those represented and required by various laws
- the provision of health and safety training
- health and safety implications of the introduction of new technologies.

r.5 The right to be provided with sufficient information:

- to enable them to carry out their functions
- on accidents that have happened, but not where:
  - an individual may be identified
  - it could prejudice the company's trading
  - the matters are the subject of litigation
  - it is against national security
  - it would contravene a prohibition imposed by law.

r.7 The right to be allowed time off to:

- receive training
- carry out a representative's functions.

r.6 The functions of a safety representative are to:

- make representations to the employer on hazards and incidents that affect employees health and safety
- raise with the employer other matters affecting the health and safety of employees
- be a contact with and receive information from HSE inspectors
- investigate potential hazards and accidents.

In addition, union-appointed safety representatives can:

- investigate complaints by employees concerning health and safety matters
- attend meetings of the safety committee
- carry out:
  - subject to giving suitable notice to, and getting the agreement of the employer, inspections of the workplace:
    * at suitable intervals
    * when there have been substantial changes in workplace
    * following notifiable accident, dangerous occurrence or disease.

Guidance on these Regulations is given in HSE Publication L87.

**4.3.2
Safety committee**

One of the most important vehicles for joint consultation is the safety committee.

To be effective safety committees need to:

- be properly constituted
- include representation from the shop floor, supervision and management
- have written terms of reference
- have as chairman someone who:
  - can run a meeting
  - is committed to high standards of health and safety
  - has authority to accept and initiate action on committee's recommendations
- work to an agenda which should be distributed at least a week before the meeting
- have an agreed procedure for raising matters, i.e. not until supervision has been given adequate time to take any necessary action and nothing has resulted
- develop a system for measuring its effectiveness by:
  - recording items raised
  - list new items raised at each meeting
  - noting number of jobs completed since last meeting
  - getting explanations for delays in completion
  - agree programmes of inspections and require a report back.

Note that it is not the function of the safety committee to bypass line management. Any safety hazard or dangerous occurrence identified must be reported to supervisors for them to initiate corrective action. This is their responsibility for which they have been given authority. No matter should be taken to the safety committee before the supervisor concerned has been given an opportunity to correct it. Only if no action results may the matter be referred to the committee.

The agenda can include an item on company affairs where the chairman can keep the committee informed of company developments and proposed actions.

Safety representatives and safety committees have a positive role to play in stimulating high levels of awareness and generating high standards of health and safety in the workplace.

**4.4
Employment
relations in health
and safety**

The term 'employment relations' includes any matters that influence the relations between employer and employee particularly as they affect conditions of employment. In that respect, health and safety are major features of employment conditions.

Safety performance can be a good indicator of the state of employee relations. Conversely, good employment relations generate high safety performance.

Foundations for good employment relations include:

- good communications – both ways
- honesty in dealing with people

- openness of approach
- clear, well understood and agreed safety rules
- well defined and understood grievance and disciplinary procedures
- discipline which, when used, must be seen to be fair
- good consultation arrangements that are seen to work
- trained and competent supervision
- visible action following complaint/request or an explanation of why not.

With the proliferation of health and safety laws and standards, questions of discipline may arise that revolved round health and safety issues. It is necessary to be aware of the recognized procedures and practices in this field to ensure speedy and effective settlement of matters. These procedures and practices stem from the vast body of employment law which exists today. That law includes the *Employment Rights Act 1996* (ERA), as amended by the *Employment Relations Act 1999* and the *Employment Act 2002*, and deals with, inter alia:

- the right of employees to a written statement of particulars of their employment
- protection for shop workers and betting shop workers working on a Sunday
- protection against suffering detriment in their employment
- suspension from work for health and safety activities
- termination of employment
- redundancy payments.

The *Employment Relations Act 1999* allows an employee to be accompanied at disciplinary and grievance hearings if he so requests.

Where disciplinary matters do arise, the Advisory Conciliation and Arbitration Service (ACAS) Code of Practice 1 on 'Disciplinary practice and procedures in employment' recommends that disciplinary procedures should:

- be in writing
- specify to whom they apply
- require that matters are dealt with quickly
- outline the disciplinary action that can be taken
- specify the levels of management that have the authority to take the various forms of disciplinary action, ensuring that immediate superiors do not have power to dismiss without reference to more senior managers
- ensure individuals are informed of complaints against them and that they are given an opportunity to state their case before decisions are reached
- give individuals the right to be accompanied by a union representative or by a colleague
- except for gross misconduct, ensure that no employee is dismissed for a first breach

- ensure cases are fully investigated before disciplinary action is taken
- ensure individuals are given an explanation for any penalty imposed
- specify appeal procedure.

In practice, should discipline become necessary, the steps outlined below form a sound procedure to follow:

1   Give an oral warning with an opportunity to improve/correct the fault complained of, offer additional training if relevant. Also, allow employee a chance to explain his side of the complaint.
2   If no improvement, give a first written warning including a statement of the possible consequences of not improving, i.e. dismissal.
3   Give a second written warning, repeating the statement given in first letter.
4   Give a final written warning including a statement about possible dismissal.
5   If no improvement, dismiss.

Other legislation affecting employment includes:

The Disabled Persons (Employment) Act 1944 – requires the employment of a certain percentage of registered disabled persons. Care must be taken that, because of their disability, their health and safety is not put at risk, especially in an emergency.

The Equal Pay Act 1970 – requires equal treatment for men and women in the same employment and doing 'like work'.

Sex Discrimination Acts 1975 and *1986* require that a person is not treated differently because of his or her sex or marital status. It applies equally to men and women. It removed restrictions on overtime and night work for women but allowed discrimination on health grounds. However, the *Employment Equality (Sex Discrimination) Regulations 2005* add further to the right to equal opportunities and treatment for men and women in employment and occupation.

Race Relations Act 1976 is similar to the Sex Discrimination Acts but is aimed at ensuring that no one will be prejudiced in employment by virtue of their race, colour, nationality, ethnic or national origins. Issues arise with Sikhs and the wearing of hard hats and employment in the food industry.

Similarly, *The Disability Discrimination Act 1995* requires that no one is discriminated against because of their disability. It has been supplemented by the *Disability Discrimination (Employment) Regulations 1996* and the Code of Good Practice on the Employment of Disabled People. Where disabled people are employed care must be taken to ensure that, because of their disability, their health and safety is not put at risk and that suitable arrangements are made for their escape in an emergency. The *Disability Discrimination Act 2005* added provisions in relation to mental illness.

The *Employment (Age) Discrimination Regulations 2006* make it unlawful to discriminate against a person because of their age. While a default retirement age of 65 is set this can be varied by agreement between the employer and the employee.

The state of employment relations reflects the culture of an organization and the attitudes, involvement and commitment of senior managers. Employees tend to react to the behaviour of those managers so an important facet of good employment relations is the setting of a good personal example by the senior people in the organization.

The health and safety of employees can be affected by the number of hours they work at a stretch and the number of days they work without a rest day. This has been the subject of an EU directive, no: 93/104/EC 'The Working Time Directive', which is incorporated into UK law through *The Working Time Regulations 1998*.

These regulations:

r.2   Define:

- night time as a period of not less than 7 hours which include the period from midnight to 5 am
- night worker as one who works for at least 3 hours of his normal daily work during night time
- rest period as a period specified in the Regulations but not including rest breaks or holidays
- working time as any time the employee is working under the employer's control, when receiving training and other times covered by a works agreement
- young worker as an employee who has reached the age of 15 years but is not yet 18 and who is over compulsory school age
- adult worker as an employee over 18 years of age
- day as a period of 24 hours beginning at midnight.

r.4   A worker's working time, including any overtime, is not to exceed an average of 48 hours averaged over 7 days for each reference period of 17 weeks.

r.5   Allow a worker, by agreement in writing, to opt out of the working time restrictions, subject to a record being kept of the hours actually worked.

r.6   Night work must not exceed an average of 8 in each 24 hour period averaged over a 17 week reference period.

r.7   When assigning work that may involve night work, employees are to be offered a free health assessment.

No young worker may work between 10 pm and 6 am unless offered a health check before starting the work.

r.8 Where the work is monotonous, workers should be given suitable rest breaks.

r.9 Employers to keep records of the hours worked and retain them for 2 years.

r.10 Daily rest periods to be allowed are:

- adult workers – 11 consecutive hours in each 24 hour period
- young workers – 12 consecutive hours in each 24 hour period.

r.11 Workers are entitled to weekly rest periods:

- adult workers, 24 consecutive hours in each 7 day period or
  2 × 24 consecutive hours in each 14 day period or
  1 × 48 consecutive hours in each 14 day period
- young workers to have 48 consecutive hours in each 7 day period

Rest periods are not to include work breaks, meal breaks, etc.

r.12 If the work period is more than 6 hours, workers are entitled to a rest break of up to 20 minutes. Young workers are entitled to 30 minutes break after 4½ hours work.

r.13 After 13 weeks employment workers are entitled to between 3 and 4 weeks annual leave which may be taken in instalments but must be taken in the leave year it is due. It must not be replaced by payment in lieu.

r.14 Deals with leave entitlement when a worker leaves during a leave year.

r.15 The dates when leave is taken must be agreed with the employer.

r.16 Workers to be paid for annual leave.

r.17 Where a worker's contract of employment allows a certain period of leave, the worker may opt for either that period or the period allowed by the Regulations but may not opt for both.

Part III (rr 18–27) covers exclusions from various requirements of these Regulations and encompasses:

r.18
- transport industry
- work at sea including fishing

- doctors in training
- the emergency services

r.19 • domestic servants

r.20 • managers and executives
- family workers
- officials at religious ceremonies

r.21 • workers who have to travel long distances to work or between jobs
- security guards and caretakers
- hospital workers
- media personnel
- utility companies
- agriculture
- where work cannot be interrupted on technical grounds

r.22 • shift workers changing shifts

r.23 • workers subject to collective or workforce agreements

r.25 • the armed forces

r.26 • young workers on board ship

r.27 • young workers undertaking work for which no adult worker is available, but subject to certain safeguards

r.24 Requires that where certain workers are required to work through their rest periods or rest breaks, compensatory rest periods must be taken

Parts IV and V deal, inter alia, with:

- enforcement
- offences
- right not to suffer detriment
- unfair dismissal
- agency workers
- police services

The originating EU directive has been amended extending the cover to all non-mobile workers with a phased application to junior doctors. The Working Time Regulations have been amended to incorporate these EU changes.

**4.5
Human factors in
health and safety**

In any activity involving human beings the effectiveness with which it is carried out depends very much on the way in which the individuals concerned look at what is to be done. Approaches can range from supreme enthusiasm that brooks no setbacks to downright indifference that looks for any reason not to do anything. The difference between these two extremes is in the attitude and manner in which the individual approaches the activity. This may be due to internal factors that are very personal to the individual or to external factors such as the circumstances or the environment in which the individual finds himself and over which he has no control. In the occupational field these factors and any others that affect the interface between employees and their work are referred to as *human factors*.

It is being increasingly recognized that many of the accidents that occur at work are a direct result of human factors, i.e. the accident has a behavioural cause rather than the failure of a mechanical part or a short-coming in a system of work.

Occupational **human factors** refer to any matters that influence a person's approach to work and the ability to carry out the job tasks. Those influences can occur at any point in the person's daily activities whether in the home, at work, at social gatherings or during leisure activities Human factors are one of the many facets of behavioural science.

This section, for simplicity and convenience, will consider the influences on human factors under four headings:

1   What human factors cover.
2   Positive factors – some factors that can improve attitudes to work.
3   Negative factors – some factors likely to increase risks at work.
4   Personal factors.

1   Human factors cover:

- the attitude of employees to their work
- relationships between employees and their work groups
- interaction between an individual and the job or work environment
- individual capabilities and fallibilities (human error)
- personal behaviour of individual
- the extent of training and instruction provided
- the design and condition of plant and equipment
- rules and systems of work – whether reasonable and acceptable.

2   Positive factors:

- a managerial environment that sets a proper safety culture
- matching the individual to the job or machine
- on-going training covering:
  - skills to carry out the work
  - knowledge of processes
  - use of work equipment
  - company plans and aspirations

- providing equipment that is:
  - safe
  - kept in good condition
  - can be adjusted to operator's capabilities of speed, size, dexterity, etc.
  - ergonomically designed
- having performance goals that are:
  - realistic
  - attainable
  - understood
  - acceptable
- discipline which is seen to work and be fair
- provision of adequate information about:
  - the job to be done
  - the company
  - work targets
- monitoring performance and communicating results
- having a system of 'feedback' to enable employee's ideas to be utilized and recognized
- ensuring agreed rules and procedures are followed.

3   Negative factors

- lack of training in tasks
- 'macho' attitude towards rules and safeguards
- ignoring or bypassing safeguards and taking short-cuts to increase take-home pay
- ignorance of what is going on
- ignorance or misunderstanding of what needs to be done
- failure to communicate or instruct properly
- bad design and layout of plant and equipment that does not take account of human limitations, both physical and mental (ergonomics)
- lack of clear direction.

4   Personal factors:

- individual attitudes to the job and work
- degree of personal motivation from the work
- whether the training received satisfies personal needs
- perception of the value of role in the organization
- ability to match demands of the job
- seeing the work as a challenge.

In their publication no: HSG48, 'Reducing error and influencing behaviour', the HSE look at a range of factors and review a number of circumstances under the headings:

- the organization
- the job
- the individual (personal factors)

with the aim of stimulating interest rather than answering questions.

Building on the positive factors and working to eliminate the negative factors will make a major contribution towards a more stimulating and safer working environment.

4.6
Insurance in health and safety

No matter how prosperous a company or how substantial its position, unexpected events can occur that can be very costly to resolve and put enormous financial strain on its resources. There are very few companies large enough to be able to absorb, within their own resources, the sort of cost involved in litigation and damages, and the almost universal approach is to spread the risk to institutions set up to handle them, the insurance companies.

Not only is it the employer who looks for recompense following an incident, but so does the employee. While the employee may be able to make a claim for damages against his employer, this is not always so and other avenues for obtaining compensation exist for him through the social security legislation.

This section deals with the sources of recompense for an employee as well as the role of insurances companies in providing sources of finance to the employer to meet unexpected demands resulting from accidents. It also looks at the influence the insurance companies have in stimulating high standards of health and safety.

4.6.1
Sources of recompense for the employee

1   From the State

(i) *Workmen's Compensation Act 1925*:
  - brought in to ensure that those injured at work received some compensation, but there were restrictions to its application

(ii) *National Insurance (Industrial Injuries) Act 1946*:
  - has largely overtaken the Workmen's Compensation Act
  - covers a broad range of benefits
  - provides payments for absences due to other than industrial causes

(iii) *Social Security (Recovery of Benefits) Act 1997*:
  - allows the State to reclaim from those paying compensation the amount of State Benefit paid

(iv) *Social Security (Industrial Injuries) (Prescribed Diseases) Regulations 1980*, as variously amended:
  - lists those conditions that attract benefit
  - requires exposure for a period of over ten years
  - limits period within which a claim must be made to five years
  - specifies industrial processes causing ill-health that can attract benefit.

2   From the employer

A claim for compensation from an employer could be on the grounds of:

- *negligence* if the harm arose because the employer failed to meet a common law duty of care, or

- *breach of statutory duty* if the harm arose because the employer failed to comply with a statutory requirement, or
- both of the above.

If a claim is successful, compensation will only be received if adequate funds are available. This is ensured by employers having Employer's Liability insurance which is a mandatory requirement of:

Employer's Liability (Compulsory Insurance) Acts 1969 and 1998

> - Requires employers to carry insurance to cover the likely cost of claims for bodily injury or disease sustained at work (EL insurance).
> - Injury or disease must 'arise out of or in the course of employment'.
> - Requires certificates of insurance to be retained for 40 years and to be kept readily available for inspection.

The 'duties of care' an employer owes to his employees has developed from earlier claims litigation and includes:

- providing a safe place of work
- providing sufficient, safe and suitable plant
- keeping plant well maintained
- providing safe systems of work
- providing adequate and competent supervision
- employing responsible people.

An injury resulting from a breach of any one of these duties of care could be grounds for a claim. (Claims for compensation are civil actions under common law.)

The employer has available to him a number of defences:

- *Volenti non fit injuria* – meaning the injured person consented to run the risk. This defence is unlikely to succeed in the present judicial atmosphere.
- *inevitable accident* – meaning that it occurred in spite of reasonable care by the employer.
- *Contributory negligence* – a partial defence attempting to put some of the blame on the claimant.
- *Res ipsa loquitur* – a plea used if the employer can prove that the accident could have occurred even if he had not been negligent.
- The action was brought outside the permitted time limit, normally three years, but in some cases (diseases) will be allowed within three years of 'date of knowledge' of the cause of the condition.

The EL insurance cover is normally £5–10 m and extends to include the legal costs as well as any compensation that has to be paid. However, if there has been a prosecution it does not cover the cost of any fine imposed.

EL policy premiums are based on declared payroll figures and cover all those included. Some necessary but voluntary work activities, such as fire fighting training, first-aid practices, etc., are undertaken outside working hours so are not included in the payroll and any injuries suffered during these voluntary activities are not covered. However, if the individuals concerned are paid a nominal honorarium, and this is included in the payroll return, they will have EL cover for those activities.

Following an accident that is likely to give rise to a claim, the employer should notify his insurers and submit a claim form.

On receiving a writ from an injured employee claiming damages:

- The writ is *not* acknowledged but sent to EL insurers.
- Insurers take over handling the claim.
- Insurers either handle the case themselves or appoint solicitors to do so.
- Copies of all information about the incident should be sent to insurers including:
  - accident reports
  - form F2508
  - photographs
  - plans of site, plant, accident site, etc.
  - medical certificates.
- Insurer's engineer or solicitor investigates.
- If claimant wants representative to visit site, refer to insurers for decision.
- Insurers negotiate with claimant's solicitor to try for out-of-court settlement since it:
  - saves costs
  - does not set precedents for future compensation levels.
- Insurer decides whether to settle claim or go to court (usually on financial grounds rather than any question of right or wrong).

4.6.2
Other insurance cover

In addition to the mandatory EL cover, it is prudent for an employer to carry insurance cover for a number of other potential risks:

1 Public liability:

- Covers claims for damages from non-employees, e.g. visitors, contractors, neighbours, members of the public.
- Covers the employer's vicarious liability for actions of his employees.

2 Fire insurance:

- Covers losses arising from a fire.
- Includes products, plant, equipment and other assets.
- May be extended to cover loss of profits as a result of a fire.
- Insurers give advice on precautions to reduce fire risks.
- Insurers may place fire precaution conditions to providing cover.

3 Plant and machinery:

- Covers the equipment itself but not injuries arising from its use.
- Insurers have specialist engineers to offer advice on good operating as well as safety techniques and practices.

- This advice may become a condition of cover.
- Insurer's engineers carry out statutory inspections within the policy.

4   Product liability:

- Provides cover against claims for injury, illness or other loss attributed to faulty product.

While the insurance industry's involvement may appear to be low key, it is able to exert an enormous influence to get compliance with required standards through:

- advice on safeguarding of machinery and plant
- advice on techniques for achieving high standards in other operations
- advice on risk management techniques
- reduction of premiums if advice is taken
- the threat to increase premiums if standards are not met, and ultimately to withdraw cover if advice is not followed.

5 Workplace safety

This part is concerned with general safety about the workplace and not with safety in relation to particular processes or machines. It covers access to and egress from as well as movement within the workplace and extends to include welfare facilities.

5.1 Workplace regulations

Requirements for safety in and about the workplace are contained in the *Workplace (Health, Safety and Welfare) Regulations 1992* and its associated Approved Code of Practice no: L24, which should be referred to for more detailed guidance.

r.3 The Regulations apply to all workplaces except:

- on a ship
- on building sites
- a workplace below ground at a mine
- on mineral extraction sites
- in transport vehicle
- on agricultural and forestry land but does apply within buildings.

r.4 Places the onus on the 'person in control', i.e. the local manager, for ensuring that the building meets requirements, is stable and is solid.

r.5 The workplace, its fixtures, fittings, built-in or attached equipment, devices, systems and safety equipment must be:

- well maintained
- kept clean
- in an efficient state
- in efficient working order
- in good repair preferably backed by a system of planned maintenance with appropriate records.

Maintenance includes:

- inspection
- adjustment

Working in confined
spaces

- lubrication
- cleaning
- all equipment, buildings, lighting, escalators, etc.

r.6　Workplace atmosphere:

- to be kept wholesome by
  - opening windows
  - wall- or roof-mounted fans
  - air-conditioning units providing either fresh air or recycled air
- in close or humid areas workers are allowed breaks in a well ventilated area
- where ventilation is necessary to protect employees, the system should:
  - be fitted with a failure alarm
  - supply more than 5–8 litres/second/worker of fresh air
  - be regularly maintained, cleaned and its performance checked
  - not cause draughts
- does not apply to:
  - work in confined spaces, i.e. vats, tanks, sewers, etc., where requirements are laid down in the Confined Spaces Regulations 1997
  - processes where extract ventilation is already required by law, i.e. working with lead, asbestos, radioactivity, dusts and under the COSHH Regulations.

r.7　Temperature during working hours:

- minimum for normal employments: 16 °C (60.8 °F)
- minimum for strenuous work: 13 °C (55.4 °F)

Note: these are minimum temperatures. Actual temperature could be higher to ensure reasonable comfort in the work area, e.g. offices may require a temperature of 20 °C (68 °F) or more.

- in high-temperature work areas such as boiler houses, furnace floors, etc.:
  - provide special cooling arrangements
- in low-temperature work areas such as refrigerators, meat stores, etc.:
  - provide warm clothing
  - a warming room facility
- in the open air or in buildings open to the atmosphere:
  - provide warm clothing in winter
  - the facility of a warming room
- no maximum working temperature is quoted only that it must give reasonable comfort
- high temperatures from sunlight may require the provision of blinds or window shading
- heaters or coolers must not release injurious or offensive fumes into the workplace
- buildings should be thermally insulated to maintain temperatures
- sufficient thermometers to be located in work areas.

r.8   Lighting:

- must be 'suitable and sufficient' (no levels are quoted but figures are given in the Chartered Institution of Building Services Engineers' (CIBSE) 'Code of interior lighting')
- wherever possible by natural light
- emergency lighting must be provided where failure of the lighting can give rise to danger, i.e. where there is no natural lighting
- particular attention should be given to lighting:
  - in rooms with display screen terminals
  - for outside roadways and footpaths
  - on building/construction sites (against glare)
  - in areas with strong shadows
- advice and information on lighting levels from CIBSE.

r.9   Housekeeping:

- the workplace, furniture and fittings must be kept clean
- walls, floors and ceilings must be kept clean
- check for build-up of dust on flat surfaces especially on building structure, roof girders, etc.
- painted walls to be washed and repainted at suitable intervals (say twelve months and seven years respectively) or after changes to the plant or building
- floors should be kept clean by sweeping or washing regularly (at least once a week)
- rubbish must not be allowed to accumulate – it can be a health risk and is a fire risk

- rubbish and waste should be put in suitable containers:
  - oil- or solvents-contaminated waste should be put into fireproof containers such as metal dustbins with lids
- spillages should be mopped up using a suitable absorbent material.

r.10 Space:

- workplaces used for the first time or modified after 1st January 1993:
  - space per employee should not be less than 11 cu. m. (388 cu. ft)
  - space above 3 m. (9 ft 10 ins) should not be considered
- for workplaces in use before 1st January 1993 and not modified:
  - space per employee should be not less than 400 cu. ft
  - space above 14 ft should not be considered
- space taken up by furniture, cupboards, cabinets, etc. may need to be deducted from the space available
- in workplaces with very low ceilings, warning notices should be displayed, and any low beams clearly marked.

r.11 Workstations:

- must be suitable for anyone likely to have to work there
- protected from the weather if reasonably possible
- have well marked emergency exit
- emergency exit routes to be kept clear
- floor kept clean and slip free
- tripping hazards should be removed
- if work platform above floor level it should have safety rail
- if work platform above 2 m. (6 ft 6 ins) a hand rail with intermediate rail and toe board is required
- work in cramped or awkward positions should be for short spells only
- work pieces and materials should be easily reached from work position.

Seating:

- wherever the work allows seating should be provided
- seats should be appropriate for the type of work with back support and foot rests
- typical types of seating:
  - in machine shops, stools
  - for production lines, stools or chairs as appropriate
  - for display screen work, chairs with adjustable height – a foot rest may also be needed
- should be kept in good condition and any damage repaired or seat replaced.

r.12 Floors should:

- be suitable for their purpose, i.e. pedestrian way, traffic route, supporting plant and materials, etc.
- not be over-loaded
- be level and smooth
- not contain potholes, irregularities or depressions likely to cause tripping hazard
- be clear of obstructions with items stowed in designated areas
- not be slippery
- where water is likely to be present have adequate means of drainage
- have separation between traffic and pedestrian ways by either hand railing, barriers or by floor markings
- have a substantial barrier around any openings or holes.

r.13 Falls from heights:

- the requirements are now contained in the *Work at Height Regulations 2005* (see Section 14.2) but are repeated here for convenience
- to be prevented if likely to cause personal injury
- precautions provided will depend on the particular work:
  - on scaffolding and platforms more than 2 m. (6 ft 6 ins) high a guard rail 1.1 m. (3 ft 6 ins) high with intermediate rail and toe board
  - guard rails are not required for a platform on builder's trestles

...provide means to prevent falls from heights ...

- an operating platform for a machine should have safety rails along its open sides
- stairs should have a hand rail on both sides whether free-standing or between walls
- fixed ladders should be fitted with safety hoops
- for work on roofs:
  * if permanent or frequent, provide fixed walkways with hand rails
  * if temporary, use crawler boards
  * provide edge protection
- when sheeting lorries either work from a special purpose gantry or if access to the top of the load is necessary, wear a safety harness attached to overhead girder or roof truss
- if access is necessary to the edge of a platform, the safety rail may be removed but only for the minimum possible time
- where access is necessary to an elevated platform for the transfer of goods, proprietary safety gates should be used.

Falling objects:

- precautions must be taken to prevent injury from falling objects:
  - by providing toe boards on scaffolding and working platforms
  - bricks and building materials on scaffold platforms should be kept in wire-mesh skips and only removed when being used
  - the provision of rubbish chutes on building sites
  - preventing access to areas below where work is going on
  - racking should be robust and stable with heavy goods stored at low level
  - goods in free-standing stacks should be 'bonded' or the height of the stack restricted (standards developed by industries for stacking their product should be followed, i.e. palletization, end-stacking of paper reels, etc.).

Access ways over vats, etc.:

- walkway must be fitted with guard rails and toe boards on both sides
- vessels containing hazardous substances should have a substantial cover
- vessel includes:
  - tanks
  - pits
  - sumps
  - silos for grain and granular materials
  - hoppers for coal, chemicals, etc.
  - kiers for boiling textile materials
  - hydropulpers.

Guard rails:

- new guard rails
  - top rail 1100 mm (3 ft 6 ins) from walkway
  - toe board 150 mm (6 ins)
  - intermediate rail half-way between top rail and toe board
- old guard rails
  - top rail 915 mm (3 ft)
  - toe board 150 mm (6 ins)
  - intermediate rail (if fitted) half-way between top rail and toe board.

r.14 Glazing:

- transparent or translucent materials used in doors, windows, walls, partitions, etc. at widths greater than 250 mm (10 ins) should be:
  - safety material:
    * polycarbonate
    * glass blocks
    * safety glass, i.e. glass that when broken does not splinter
    * annealed glass of appropriate thickness
    * Georgian wired glass
- large plain sheets of glazing that form a door of an accessway should be marked to make them apparent.

r.15 Skylights and ventilators:

- must be capable of being opened safely
- when open not project and cause a danger to passers-by.

Windows:

- capable of being opened safely
- when open not project and cause a danger to passers-by.

r.16 Windows must be capable of being cleaned safely (possibly using special equipment).

r.17 Traffic routes:

- include both pedestrian and vehicular traffic
- pedestrian routes should:
  - wherever possible be separated from vehicular routes
  - be of adequate width
  - have non-slip surface
  - accommodate the disabled including wheelchairs

- have wheelchair ramps where steps occur
- be clearly identified by signs and/or floor marking
- kept clear of obstructions
- have protective guard rails at blind exits from buildings
- vehicular routes should:
  - be wide enough to accommodate the vehicles that use them
  - be of suitable construction to support the vehicles
  - be kept well maintained without potholes
  - be properly drained
  - have speed limits with speed humps
  - be clearly marked
  - have signs and road markings that comply with road traffic signs and markings
  - have hazards such as limited headroom or width clearly identified with advance warnings
  - avoid hazards such as the edge of pits or trenches, susceptible structures such as cast iron support columns, power cable poles, chemical filler pipes, etc.
  - allow for passing points and parking
  - have adequate manoeuvring space by loading bays
  - have safety refuges in loading bays
  - be provided with pedestrian crossing points.

r.18 Doors and gates:

- all doors and gates should be:
  - of proper construction for their function and size
  - fitted with proper latches and closing devices
  - if opening both ways have a viewing panel
- power operated doors and gates should:
  - have means to prevent anyone being trapped such as:
    * sensitive edge trip that causes the door to open
    * closing force low enough not to cause injury
    * manually operated control by hold-down button, release of which causes the door to open
    * if power fails:
      - be capable of being opened manually or
      - open automatically on power failure
      - except lift gates which must remain closed and only open at landings
      - have a local isolating switch for use in emergencies
- sliding doors should:
  - be fitted with a device, such as a fixed rail over the running wheels, to prevent them leaving the track
  - have means to guide the lower edge
- vertical rising doors should have an anti-fallback device, i.e. be counter-balanced or fitted with a ratchet

- fire doors are intended to retard the spread of fire and should:
  - be installed at fire exits and along fire escape routes
  - be self-closing
  - be capable of retarding fire for a specified time:
    * most common is $\frac{1}{2}$-hour fire break-door
- fire doors on traffic routes through fire-break walls may be held open by a device such as a fusible link or magnetic device activated by the fire alarm that release the door automatically in the event of a fire
- smoke doors fitted in corridors should:
  - be of the $\frac{1}{2}$-hour fire-break type
  - close to a rebate or, if double swing, have edge seals to contain smoke
  - be self closing
  - be fitted with a viewing window.

r.19 Escalators and travelators should:

- be fitted with devices to prevent trapping between:
  - the treads and end-comb
  - treads and side-plates
- have readily accessible and easily identifiable emergency stops.

r.20 Toilets should be:

- easily accessible and clearly identified
- under cover
- separate for men and women
- private
- arranged so that urinals are not visible from outside men's toilet when door open
- fitted with obscured glass in the windows or curtained
- kept clean and tidy
- well ventilated and lit
- provided with a ventilated space between toilet and work/public area
- provided with facilities for washing hands and drying them
- suitable for the needs of the disabled and wheelchair bound.

WC cubicles should:

- have locks on the inside of the doors
- be large enough to give privacy
- be provided with toilet paper.

Number of toilet facilities to be provided:

- in post-1992 work premises:

max number on premises	no. of WCs	washbasins
1–5	1	1
6–25	2	2
26–50	3	3
51–75	4	4
76–100*	5	5

* thereafter one WC and one washbasin per twenty-five persons

- if premises occupied by men only:

max number on premises	no. of WCs	no. of urinals
1–15	1	1
16–30	2	1
31–45	2	2
46–60	3	2
61–75	3	3
76–90	4	3
91–100*	4	4

* thereafter one WC and one urinal per fifty men

- in pre-1993 premises:
  - factories:
    * 1 WC per 25 women employed
    * 1 WC per 25 men employed (urinals are additional but no number given)
    * in factories employing over 100 men the ratio of facilities for men reduces
  - offices (for both sexes):

number employed	no. of WCs
1–15	1
16–30	2
31–50	3
51–75	4
76–100*	5

  * thereafter one WC per twenty-five employees
  - if urinals are provided for men, the number of WCs reduces by one.

r.21 Washing facilities should:

- be located in or near toilet facilities and/or changing rooms
- be easily accessible
- be supplied with hot and cold water
- be provided with soap and towels or other method of drying
- be well ventilated and lit

- be kept clean and tidy
- have separate facilities for men and women unless washing is restricted to hands, arms and face only
- as a guide, have one wash basin per WC – if work is dirty there should be one washbasin per ten employees
- if showers are provided they should:
  - enjoy the same privacy as toilets
  - have thermostatic control on hot water supply.

r.22 Drinking water supply must be:

- adequate
- potable
- readily and easily accessible
- clearly marked and suitably positioned
- provided with 'Drinking water' or similar label on tap
- drinking fountain or provided with cups or drinking vessels
- in a special bottled water dispensor.

r.23 Clothes accommodation:

- for storage of non-work clothes
- separate storage of work clothes
- clothes storage must be secure
- provide facilities for drying wet clothes including low-level heating.

r.24 Changing rooms should be:

- provided for changing clothes
- separate for men and women
- provided with:
  - benches or seats
  - storage facilities
  - heating
  - washing facilities if appropriate
- large enough to accommodate the number likely to be changing at any one time.

r.25 Rest rooms should:

- be available or a rest area
- be private
- have facilities for lying down
- have curtains or other means to subdue the light
- be adjacent to first aid facilities and toilets
- include facilities for pregnant women and nursing mothers to rest

- ban smoking
- be kept clean and tidy preferably under control of nurse or first aider.

Eating facilities (small facilities and not canteens with paid staff) should:

- be separate from work area
- include the provision of hot water or hot and cold drinks dispensers
- be provided with suitable tables and chairs
- be provided with a sink or means for washing eating utensils
- be kept clean and tidy (it may be subject to hygiene inspections)
- contain a refrigerator for keeping foods fresh
- have means for heating food – small oven or microwave
- ban smoking (this is now a statutory requirement)
- have facilities to accommodate the disabled and wheelchair bound.

Any signs provided to indicate the location of facilities and escape routes should comply with the *Health and Safety (Safety Signs and Signals) Regulations 1996*.

These requirements demand only what is reasonable for a decent quality of working life.

**5.2
Office safety**

While offices are workplaces and as such have to comply with the requirements of the *Workplace (Health, Safety and Welfare) Regulations 1992* as summarized in Section 5.1 they do contain a number of hazards that are specific to them and for which there are particular precautions.

The following is a list of some of the hazards met in offices and typical precautions that can be taken.

Location	Hazard	Precautions
Floor	Polished and slippery	Use non-slip polishes Cover floor with non-slip mats
	Water or greasy spillages	Clean up as soon as found
	Threadbare carpets	Tripping hazard; remove or replace
	Man-made fibre carpets	Generate static electricity when walked on particularly with plastic-soled shoes Sprinkle carpet with water or use anti-static spray

Filing cabinets	Overloaded drawers	If top drawer is heavy, cabinet can topple when drawer opened Keep heavy loads to lower drawer Attach cabinet to wall or to back of other cabinet
	Drawers left open	Form tripping hazard and obstruction Ensure drawers shut when not in use
Shelves	At high level	Reaching up to place items Provide suitable steps
Electrical equipment	Loose leads across the floor	Remove or enclose in special-purpose ramped floor conduit
	Condition of leads, etc.	Regular inspection of plugs, sockets, leads and appliances by qualified electrician and replacement of faulty items
	Private equipment	Use of unofficial equipment, i.e. kettles, radios, drills, heaters, etc. should be prohibited
	Repair of equipment	Carried out by qualified electrician only Equipment must be isolated from electrical supply
Photo-copiers	Clearing jams	Power must be switched off Usually happens automatically when cabinet door opened Follow maker's instructions Do not use pointed instruments to probe
	Fumes	Ozone may be generated (WEL – 0.2 ppm) Ensure area is well ventilated
Printing machines	In-running nips	Ensure operator fully trained Nip guard is in position
	Use of solvents and inks	Keep quantity to minimum Use non-spill containers Provide good ventilation Provide protective gloves Provide washing facilities nearby Mop up any spillages
Fire precautions	Smoking	Prohibit Provide special designated smoking areas with facilities for extinguishing dog-ends

	Smoke and fire doors	Keep closed If volume of traffic requires them to be open, retain in open position by automatic device which releases when fire alarm activated
	Extinguishers	Correct type for possible type of fire Regularly inspected and maintained Properly mounted 1 m. from floor level
	Fire escape routes	Clearly marked Kept clear at all times Known to all staff
	Fire-exit doors	Must be kept unlocked at all times persons are on premises If for security reasons must be locked, use panic bar or break-glass bolt
Wastepaper baskets	Contents	Restrict to paper only Risk of fire from surreptitious dog-ends; Other items such as broken glass or razor blades to be parcelled separately and put to one side for cleaner to collect Metal or non-flammable construction
Display screen equipment (DSE or VDU)	Carry out risk assessment	See Section 3.4
	Work station	Ergonomic layout – see Section 10.3
	Noise from some printers	Provide acoustic hood
	Radiation emissions	Very low-level and no risk to health or pregnancy
	Rest pauses	Regular breaks from work station – move around, especially important for pregnant women
Furniture	Condition	Well maintained Suitable for the work Remove sharp corners and splinters

Ventilation	Natural ventilation	Openable windows
	Forced ventilation	Window or wall-mounted fans – blades guarded
		Free-standing circulating fans – blades guarded
	Draughts	Adjust means of ventilation to reduce to minimum
	Smokers	Ban
Lighting	Adequate level	Properly designed artificial lighting system
		Eliminate glare
		Use natural light wherever possible
		If glare from sunlight, provide suitable blinds

Although offices are generally considered to be safe they still contain many hazards that can cause serious injury. The majority of office accidents are of behavioural origin and can be avoided by taking care, being considerate for others and treating equipment with respect.

**5.3
Workplace safety
signs and signals**

The use of the correct safety signs in the workplace can:

- reinforce safety instructions and rules
- give information on risks and precautions to be taken.

There are three separate sets of safety signs that may be used in the workplace:

1 Safety signs used to give information in the course of normal work.
2 Hazard warning signs identify the hazards of various substances and are required to be included as part of the labelling of hazardous substances. See Sections 11.2; 11.3 and 11.4.
3 Hazchem Boards to give the emergency services warning of the chemical nature of substances that may be involved in fires or road incidents. For transport vehicles they are supported by a Tremcard held by the driver. See Section 11.8.

For workplace safety signs the type, shape and colour of the signs are specified in *The Health and Safety (Safety Signs and Signals) Regulations 1996* which apply only to workplaces. Signs concerning the identification of hazardous and dangerous substances are dealt with in Sections 11.2, 11.4 and 11.8.

The Regulations require:

r.4 Where, after a risk assessment, residual risks remain suitable warning signs must be posted to indicate the nature of the risk and precautions to be taken.
  Suitable fire safety signs to be posted to meet the needs identified by the fire risk assessment or the recommendations of a Fire Authority (additional signs may be necessary on the recommendations of fire insurers).

Any signs provided must be kept maintained and in place.

Where a risk arises from road traffic, a sign prescribed by the *Road Traffic Regulation Act 1984* may be used if it is appropriate.

Where hand signals are used they should comply with the signals illustrated in either:

- part IX of schedule (see 5.3.2 below) or
- BSS 6736: 1986 Hand signals for agricultural operations, or
- BSS 7121: 1989 Code of Practice for the safe use of cranes, or
- Appendix C of the Fire Training Manual

r.5 Employers must provide for their employees:

- comprehensive information on the interpretation of the signs
- training in
  - the meaning of the signs
  - action they require to be taken.

There are four colours which have specific meanings:

5.3.1 Safety sign colours

Colour	Meaning	Information
Red	Prohibition sign	Action shown must NOT be carried out
	Danger sign	Shutdown, evacuate, emergency devices have operated, stop actions
	Fire fighting equipment	Identification of equipment and its location
Yellow	Warning sign	Take care, take precautions, proceed cautiously
Blue	Mandatory sign	Instruction MUST be followed Equipment indicated MUST be worn
Green	Safety information sign	Emergency escape routes, location of first aid post

The signs associated with these colours are of a particular design and shape:

Prohibition sign
- round shape
- black pictogram on white ground
- red edging and diagonal line

Fire fighting sign
- rectangular or square shape
- white pictogram on red ground

Warning sign
- triangular shape
- black pictogram on yellow ground
- black edging

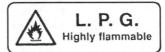

Mandatory sign
- round shape
- white pictogram on blue ground

Safety information sign
- rectangular or square shape
- white pictogram on green ground
- can be to BS 5499 pt 1: 1990 or to EU Directive 92/58EEC but must be consistent throughout workplace

Detailed information on the material and size of these signs can be obtained from reputable sign manufacturers who will also survey premises and advise on number and location of signs.

5.3.2
Safety signals

These are hand signals used in handling, lifting and other operations where the plant operator cannot see what is happening to the work piece.

In any such operation:

- there should be only one person (banksman) instructing the operator
- both the banksman and the operator should have the same understanding of the signals which should be to BS 7121: 1989
- the banksman should be clearly visible to the operator at all times (or in direct telephone/radio contact).

Typical hand signals include:

Stop

Emergency Stop

Clench and unclench fingers to signal 'take the strain' or 'inch the load'

Hoist

Lower

Slew in Direction Indicated

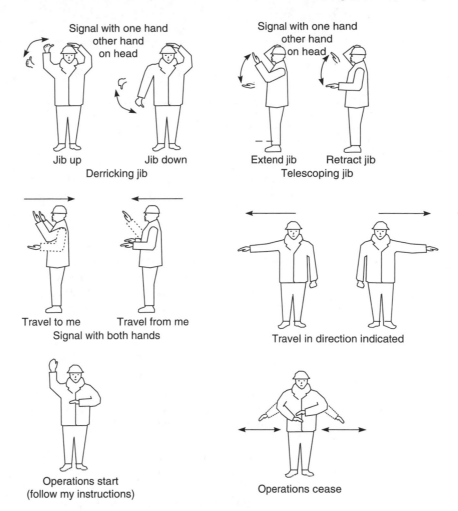

Signal with one hand
other hand
on head

Jib up   Jib down
Derricking jib

Signal with one hand
other hand
on head

Extend jib   Retract jib
Telescoping jib

Travel to me   Travel from me
Signal with both hands

Travel in direction indicated

Operations start
(follow my instructions)

Operations cease

Safety signs and signals are there to inform and make work operations safer, but they will only be effective if the workpeople know what they mean and use them.

Information and advice

Information, with its attendant need for communication, is the lubricant that ensures the smooth running of an organization. This applies very much in health and safety where, without it the motivation and enthusiasm would very quickly dry up and safety activities grind to a halt.

But it is essential that the information, when provided, is well founded with a sound professional backing.

This chapter looks at sources of safety advice and information, the transfer of information by way of reports and how information can be stored so that it can be easily retrieved.

6.1
Safety advice

Under the *Management of Health and Safety at Work Regulations 1999*, all employers should have available to them a source of competent health and safety advice. How this is achieved will depend on the size and organization of the particular employment but it can be achieved in a number of ways.

Safety advice can be provided by:

- full-time safety adviser(s)
  - normally only where:
    * the concern is large enough to support them
    * the degree of risk warrants it
- part-time appointment
  - part of a dual role for an existing employee
  - appointment of a part-time consultant
- retained consultant
  - retained to provide:
    * an agreed number of days advice
    * advice on demand at an agreed fee.

The employer must ensure that the person appointed is competent in relation to the advice to be given.

Competence can be shown by:

- for a full-time safety adviser and a consultant:
  - being qualified in health and safety:
    * having a university degree in occupational health and safety, or
    * having passed Part II of the NEBOSH Diploma examination (NVQ level 4)
  - **and** having a number of years practical experience in the field
  - being a full (corporate) member of IOSH

- being a Registered Safety Practitioner (RSP) with IOSH
- being a registered MIIRSM with the British Safety Council (BSC).
- for a part-time appointment:
  - in high-risk premises:
    * being qualified to the same level as full time appointee
    * having passed Part 1 of the NEBOSH Diploma examination (NVQ level 3)
    * having some years practical health and safety experience
  - in lesser-risk areas
    * having passed the NEBOSH National General Certificate examination
    * being familiar with the work area
  - in low-risk premises
    * having passed the NEBOSH National General Certificate examination
    * having obtained the IOSH 'Managing Safely' certificate
    * being familiar with the work area.

Where a professional consultancy is selected:

- the employer should check that they have:
  - the necessary degree of knowledge
  - the necessary degree of expertise
  - sufficient facilities to provide the appropriate level of advice.

Whichever method is used the employer is required to:

- give to the safety adviser information on particular hazards or factors involved in the work
- ensure that suitable facilities are available to enable the safety adviser to carry out any necessary investigation
- allow adequate time for carrying out the safety function
- if more than one person appointed, to ensure they co-operate.

HSE inspectors will give advice if requested but you can never tell where advice stops and inspection begins.

The fact of appointing a safety adviser does not absolve employers from their statutory responsibilities – the role of the safety adviser is to give employers advice that will assist them in making decisions on how to meet their statutory obligations. However, it is common for employers to delegate the *performance* of their obligations to a manager or other appointee.

The employer is required to consult with safety representatives on the appointment of safety advisers.

6.2
Sources of information

In the course of their daily work, managers and safety advisers need to obtain information about a whole range of subjects. It is not realistic to expect them to have all the information they need at their fingertips but they should know where to go to get that information. This section looks at some of the sources of information available to managers and safety advisers.

It is convenient to consider this in two parts; written information and oral information.

**6.2.1
Sources of written
information**

Laws:

- Acts (statutes), Regulations and Orders (statutory instruments)
- all part of statute law
- copies obtainable from The Stationery Office Ltd, PO Box 276, London SW9 5DT; Tel: 0870 600 5522; Fax: 0870 600 5533 or e-mail: book.orders@ tso.co.uk
- copies held by public reference libraries
- catalogues of statutes held by reference libraries
- state the obligations to be met.

HSC/E* publications:

(Copies obtainable from HSE Books, PO Box 1999, Sudbury, Suffolk CO10 2WA. Tel: 017 8788 1165; Fax: 017 8731 3995; e-mail: hsebooks@prolog. uk.com; or online at www.hsebooks.co.uk

- Approved Codes of Practice
  - Codes of Practice approved by the Minister
  - set standards to give compliance but recognize that other methods may be equally effective
  - quasi-legal status and can be used in court to show how obligations can be met
  - contents included in HSE's L series and HSG series of publications
- Health and Safety Regulations booklets – HSR series
  - give interpretations of regulations and how to comply
  - being superseded by Legal booklets, L series
- Health and Safety Legal booklets – L series
  - give guidance on the interpretation of and compliance with specific pieces of health and safety legislation
  - may include an Approved Code of Practice
- Health and Safety Guidance booklets – HSG series
  - give guidance on techniques and practices for a range of activities and situations to achieve compliance with statutory requirements
  - no legal recognition but may be referred to by inspectors to indicate means of compliance
  - may include an Approved Code of Practice
- Guidance Notes – five different series of guidance on specific subject topics:
  - Chemical Safety (CS series)
  - Environmental Hygiene (EH series)
  - General Series (GS series)

* The above HSC/E guidance publications are listed in the HSE Books 'Catalogue' obtainable free from HSE Books. This price list also contains information on IT services offered by HSE. The free pamphlets are listed in 'List of Free Publications' available free from HSE Books.

- Medical Series (MS series)
- Plant and Machinery (PM series)
- a range of free pamphlets giving advice on a wide range of occupational safety and health issues.

Further information on current health and safety issues can be obtained from:

- HSC Newsletter, published bimonthly and contains information on latest health and safety developments in the UK
- **hse**direct, an authoritative electronic database updated daily, subscription only, contact www.hsedirect.com
- a databased bibliographic reference via HSELINE on subscription
- an information service via HSE Infoline – Tel: 0845 345 0055.

Other sources of published information include:

- British Standards Institution (020 8996 9001 or orders @ bsi-global.com)
  - publishes British Standards – identified by number BS XX
  - publishes EU Harmonised Standards in UK – identified by number BS EN XX
  - published international standards, identified as BSI ISO xxxxx for mechanical standards or BSI IEC xxxx for electrical standards
  - EN, ISO and IEC standards take precedence over BSI standards
  - conforming to the appropriate standard published by BSI is deemed to give compliance with statutory requirements.

- Books
  - *Health and Safety* by Redgrave, Fife and Machin, Elsevier – the 'bible' of health and safety law
  - *Safety at Work* by John Ridley and John Channing – 'authoritative guide to health and safety', Elsevier
  - *Croner's Health and Safety Manager*, loose-leaf information up-dated three times a year
  - *Tolley's Health and Safety at Work*, loose-leaf information with regular up-dating
  - *Health, Safety and Environment Bulletin*, Industrial Relations Services – tends to deal with the legal aspects, useful for case law and Tribunal decisions
  - *Hazards at Work*, TUC – useful book giving sound advice but tends to be from the employee's point of view.

- Journals of the professional institutions:
  - IOSH   The Safety and Health Practitioner
  - BSC   Safety Management
  - BOHS   The Annals of Occupational Hygiene

**6.2.2
Sources of oral
information**

Safety advisers are generally happy to discuss safety problems on the phone subject to there being no commercial implications. Employer/ employee organizations and the national safety organizations tend to give advice only to their members. Safety consultants are likely to charge for any advice.

- Safety adviser
  - should be able to give information pertinent to the particular work situation

- Safety representative
  - can often give information about work hazards and solutions

- HSE inspector
  - able to give authoritative advice on a range of occupational safety related problems but if asked to call may use the opportunity to carry out an inspection of the premises

- Employer's organization
  - many have a safety department able to give advice pertinent to the particular industry

- Employee's organization
  - many have a safety section but may restrict advice to members; also advice may be strongly angled towards the employee

- The Royal Society for the Prevention of Accidents (RoSPA)
  - country's longest established safety organization
  - covers safety in all human activities
  - can give occupational safety advice but may restrict it to members

- The British Safety Council
  - an independent safety organization
  - gives advice but only to members

- Manufacturers and suppliers
  - have legal obligation to provide customers with safety information about their products
  - usually willing to assist with safety problems affecting their products

- Safety consultants
  - should be professionally qualified but credentials should be checked before using
  - can give detailed technical advice in respect of specific problems
  - can be very expensive.

| 6.3 Report writing | Reports are a two-edged sword – they can be used to get a message through to the recipient (manager) or be a means whereby the recipient finds things out about what is going on. |

To be effective, reports must:

- be properly constructed
- be structured
- present the information in a logical manner
- be concise while covering the subject
- readable with good grammatical construction
- use acceptable words not jargon
- carry the reader to the desired conclusion.

The stages in preparing a report are:

- gather information and facts
- analyse the facts
- organize the data
- put pen to paper.

Points to be kept in mind when preparing a report:

- why is it needed?
- the object of the report
- the audience it is aimed at
- what it should cover and not cover
- whether it should be general or specific and detailed
- need for the text of the report to 'flow'
- possible contents
- how it is to be structured
- identification of headings and sections – numbering
- value of diagrams, graphs and illustrations.

Suggested make-up of a report:

- title page with author's name and signature, date of issue and distribution
- summary
- list of contents
- introduction giving background information on the subject
- description of the investigation/incident/circumstances
- findings – recording of information obtained
- analysis of findings
- conclusions drawn from findings
- recommendations
- appendices/annexes of supporting material such as photographs, sketches, diagrams, tables, etc.

Before finalizing a report:

- read it through
- ask yourself:
  - does this add anything to the subject of the report? If not, discard it
  - is that what I meant to say? If not, discard or change it to say what you meant

- can that be interpreted another way? If so, rewrite it
- will that be clear to the reader? If doubtful, rewrite less ambiguously
- is it to the point? If not, discard it.

Presentation of report:

- important to attract attention
- in acetates or special covers
- title displayed on or through cover
- binding – ring/comb/staple/perfect (glued)/sewn
- distribute with covering letter.

Remember the recipient of the report is a busy person and to keep his/her attention follow the KISS principle – **K**eep **I**t **S**hort and **S**imple.
Useful references:

- *The complete plain words* by Sir Ernest Gowers, Penguin Books
- *A Guide to Report Writing* (booklet), the Industrial Society.

6.4
Data storage and retrieval

There is no sense in collecting data if you are not going to use it, and if you do collect it how are you going to store it so you can find it when you want it? Whatever system is used must match identified needs and integrate with existing data storage and retrieval systems.

6.4.1
Reasons for keeping health and safety data

- Statutory
  - accident reports
  - reports of statutory inspections of pressure systems, lifting equipment, etc.
  - risk assessments required under various Regulations
  - chemical data sheets under COSHH
  - training records
  - Fire Certificate (if issued).

- For use in company
  - the company's safety policy
  - reports of accident investigations
  - reports of safety inspections
  - minutes of safety committee meetings
  - information on health and safety items and products
  - copies of health and safety laws
  - personnel records
  - risk assessments.

6.4.2
Methods of storing data

- By diary
  - simple risk assessments
  - notes of minor and non-reportable accident.

- By hard copy
  - reports of statutory examinations
  - chemical data sheets
  - copies of accident report forms F 2508 and F 2508A
  - HSE Codes of Practice, guidance notes and advisory publications
  - copies of Acts and Regulations.

- On disk
  - scheme for examination of pressure systems
  - safety information (HSE's Infoline: HSELINE: hsedirect; etc.)
  - accident data and analysis.

6.4.3 Information retrieval

- From diary
  - by memory.

- Hard copy
  - from particular file.

- From disk
  - by using search or interrogation facility.

6.4.4 Systems for identifying stored data

- Hard copy by:
  - subject matter
  - material
  - numerical referencing system
  - alphabetic referencing system
  - haphazard system (memory).

- On disk by:
  - subject heading
  - keywork.

This may look like a lot of paper or office work but it is necessary if benefit is to be gained from past experiences and also, in certain cases, as proof of actions taken should queries arise in the future.

7   Accidents

Accidents occur – they do not *happen*; they are *caused* – by shortcomings on the part of the employer, the employee or both. The results can be traumatic for both: for the employee in the way an injury can affect him personally, his family and their quality of life; for the employer through lost production, time spent on investigations and at worst, the cost of a legal action.

Accident prevention aims to reduce the chances of such an incident to the absolute minimum.

7.1 Principles of accident prevention

Objectives

1 To prevent accidents happening.
2 If accidents do occur, to prevent recurrence.

Procedure
- Identify hazards.
- Eliminate hazards.
- If hazards cannot be eliminated, reduce hazard to minimum.
- Assess residual risk.
- Control residual risk.

Identification of potential hazards
- Pre-accident
  - risk assessment (see Section 3.4)
  - safety inspections
- Post-accident
  - accident investigation (see Section 7.2)
- Near miss
  - implement near-miss reporting procedure.

Definitions
Hazard – something with the potential to cause harm
Risk – the probability that the harm will occur and the severity of the resulting damage
Accident – an unforeseen event causing injury or damage
Near miss – an event that narrowly missed causing injury or damage.

**7.1.1
Heinrich's domino
theory**

An accident is not a single event; it is the result of a series of linked causes.

The dominoes in Figure 7.1 represent the sequence of causes (events or situations) leading to an accident that results in injury or damage. If a domino falls it will knock over the rest until eventually the final domino falls, i.e. an accident happens. If any of those dominoes (causes) is missing, i.e. the correct safety action has been taken, there will be no accident.

Typical examples of these causes are:

- Work situation
  - inadequate management control
  - lack of suitable standards
  - failure to comply with standards
  - faulty or inadequate equipment or workplace

- Fault of person
  - lack of skill or knowledge
  - physical or mental problems
  - lack of or misplaced motivation
  - inattention

- Unsafe act
  - not following agreed methods of work
  - taking short cuts
  - removing or not using safety equipment

- Accident
  - the unexpected event
  - contact with dangerous machinery or electricity
  - falls
  - being struck by equipment or falling materials, etc.

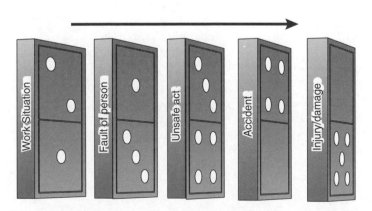

Figure 7.1 Heinrich's
domino theory

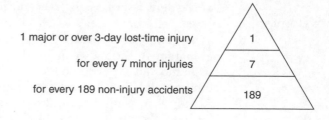

1 major or over 3-day lost-time injury
for every 7 minor injuries
for every 189 non-injury accidents

Figure 7.2 The accident triangle (from HSG65)

Resulting in:

- Injury/damage
  - to employee
    * pain and suffering
    * loss of earnings
    * loss of quality of life
  - to employer
    * damage to plant
    * compensation payments
    * loss of production
    * possible prosecution.

7.1.2 Practical accident-prevention techniques

- Near misses
  - encourage the reporting of near misses
  - investigating them can prevent serious injuries
  - need for a 'no-blame' culture

- Hazard identification
  - by inspection
  - through safety tours, inspections, etc.
  - reports from operators
  - reports in technical journals

- Hazard elimination
  - by engineering means
  - changing plant
  - changing materials
  - changing the process

- Reduction of hazard
  - engineering means, modify equipment
  - provision of guards
  - provision of PPE

- Assessment of residual risk
  - see Section 3.4

- Control of residual risk
  - engineering means – alarms, trips, etc.
  - safe systems of work
  - training of employees.

**7.1.3
Management
techniques**

- Commitment
  - implement incident-reporting procedure and monitor it
  - manager seen to be involved in and committed to preventing accidents
  - effective safety policy
  - establishment of suitable formal organization with stated safety responsibilities
  - development of appropriate rules, standards and safe systems of work
  - ensuring good communications
  - have effective consultative procedures/committees
  - use of job-safety analysis
  - monitor performance and follow up shortcomings
  - include safety from design stage
  - quality assurance and safety – BS ISO 9001 registration.

**7.1.4
Role of engineering**

- Knowledge
  - of hazards and engineering techniques to overcome them
  - of techniques for guarding machinery
  - of operator working methods

- Maintenance
  - precautions in high-risk area
  - need for operatives to be skilled
  - technical training
  - safe systems of work/permits-to-work

- Plant and equipment
  - design or adapt to suit the operator
  - maintain in good repair
  - preventative maintenance.

**7.1.5
Costs of failure**

- Direct costs
  - sick pay
  - repairs to damaged plant
  - loss of production
  - increased insurance costs

- Indirect costs
  - cost of investigation
  - loss of goodwill and/or image in the community
  - hiring and training replacement staff.

**7.2
Accident
investigation**

Objectives

- To determine the cause so recurrence can be prevented.
- Not to apportion blame.
- To obtain information for reporting to enforcing authority.
- To obtain information for insurers either:
  - in respect of a claim for pain and suffering by injured person to assist settlement or to resist litigation
  - to claim for damage to plant, equipment, etc.
- To obtain information for other statutory agencies, i.e. Social Security benefits.

Cause of accident – the events or circumstances that led to the incident that caused the injury or damage.

- Immediate cause – the part or item that actually caused the injury or damage.
- Root cause – the action or activity that resulted in contact with the immediate cause. Root-cause analysis involves checking on the sequence of events and decisions that led to the accident and identifying the often remote action that triggered that sequence of events.

Cause of injury or damage

- the action or process that occasioned the actual injury or damage.

Investigation

- By whom?
  - initially supervisor, who informs safety adviser
  - safety representative – note their rights (see Section 4.3)
  - safety adviser
  - if a claim against employer likely or has been made, insurance surveyor/engineer
  - if the injury or incident has to be reported to enforcing authority, enforcing authority inspector
  - if a fatality, police.

- When?
  - immediately injured person returns from first aid or is removed for medical treatment
  - before site of accident has been disturbed.

- Procedure
  - do not disturb the site until all the relevant details have been noted
  - view site and note significant details
  - take photographs

- measure up relevant parts and areas
- check condition of plant and equipment – arrange for tests if necessary
- interview witnesses
  * ideally alone but may have representative present if requested
  * point out that the object of investigation is to discover the cause of accident
  * evidence must be direct not hearsay
- check records of training given to injured employee
- interview injured person as soon as possible but do not cause distress
- analyse information and prepare report
- if claim has been entered, insurer's agent will want to investigate and interview witnesses but cannot interview claimant
- if enforcing authority inspector investigates, may take statements from witnesses including injured person
- in case of fatality, police investigate to determine cause of death and if there has been any criminal action.

- Inquiry
  - if inquiry held, purpose must be clearly stated, i.e. to determine the cause of the accident
  - report of inquiry available to both employer and employee, i.e. is 'discoverable' in the event of a claim
  - if object of inquiry is to resist a claim, this must be clearly stated and understood by all those involved, when notes and report could be 'privileged'.

- Information to collect
  - details of site – owner, address, department/section/shop
  - process or operation concerned including details of any plant involved
  - date, time of accident
  - personal details of injured person (possibly from personnel records)
  - information on training given to injured person
  - job being carried out at time of accident
    * was it authorized?
    * was correct procedure being followed?
    * were guards in place?
    * etc.
  - details of injury received.

- Report
  - see Section 6.3
  - analyse results of investigation and information obtained
  - prepare report setting out circumstances of accident and possible causes
  - make recommendations to prevent recurrence.

7.3
Accident reporting

There are three reasons for reporting accidents

1 Statutory obligation
2 Insurance claim
3 In-house accident prevention.

7.3.1
Statutory reporting

This is required by the *Reporting of Injuries, Diseases and Dangerous Occurrences Regulations 1995*, which imposes requirements relating to the occurrence of:

- fatalities
- major injuries
- injury to other than employee requiring removal to hospital
- certain diseases caused by listed processes
- injury or death associated with the use of flammable gas
- one of the listed dangerous occurrences.

Injuries to be reported

- Major injuries:
  - any fracture other than to finger, thumb or toes
  - any amputation
  - dislocation of shoulder, hip, knee or spine
  - loss of sight – whether temporary or permanent
  - a chemical or hot metal burn to the eye or any penetrating injury to the eye
  - electrical injuries causing burns, shock and loss of consciousness
  - any other injury causing hypothermia, heat-induced illness or unconsciousness, requiring resuscitation or admittance to hospital for more than twenty-four hours
  - loss of consciousness due to lack of oxygen
  - decompression sickness requiring immediate medical attention
  - acute illness needing medical attention or loss of consciousness from a chemical substance
  - acute illness from pathogen or infected material
  - any injury that results in detention in hospital for more than twenty-four hours.

- When?

> - major injuries
>   forthwith by quickest possible means and follow with a written
>   report within ten days
> - diseases
>   on receipt of written diagnosis from a doctor of one of the diseases
>   listed in the Regulations
> - flammable gas incident
>   on receipt of notification of an injury from flammable gas followed
>   by a written report within fourteen days
> - dangerous occurrence
>   forthwith by quickest means and follow with a written report
>   within ten days.

- How?
  - for injuries, gas incidents and dangerous occurrences
    form 2508
  - for diseases
    form 2508A.

- To Whom:
  The Incident Contact Centre by:
  - phone   0845 300 9923 (8.30 am – 5.00 pm)
  - fax   0845 300 9924 (any time)
  - internet   www.riddor.gov.uk (any time)
  - e-mail   riddor@natbrit.com (any time)
  - post   Incident Contact Centre,
                   Caerphilly Business Park,
                   Caerphilly CF83 3GG

  Or direct to the local office of the HSE or LA by:
  - phone
  - post

- What to report:
  - the information requested on:
    * form F 2508 for injuries, gas incidents and dangerous
      occurrences
    * form F 2508A for diseases
  These forms can be photocopied for faxing or imaged for e-mailing.

- By whom?

> - employer
> - person in control of the work
> - the manager of a mine
> - the owner of a quarry
> - the owner of a tip
> - the owner of a pipe line
> - if from materials being carried, the owner of the lorry.

Statutory records and accident reports must be kept readily available for inspection and retained for a period of at least three years.

Under the social security legislation all companies with 10 or more employees are required to keep an accident book in which the occurrence of all accidents should be recorded and which must be retained for 3 years after the date of the last entry. Because the original accident book contained personal data about individuals it has been replaced by a new version that is data protection compliant. This new version only must now be used.

Note: benefit under the current Social Security Act may be claimed for certain diseases and injuries. These claims are initiated by the claimant but the employer may be required to confirm the circumstances of the incident giving rise to the condition:

- for injuries – form BI 76
- for diseases – form BI 76D.

7.3.2 Reporting to insurers

Where, following an injury-causing accident, it appears likely that the injured person will claim damages, employers may wish to make a claim under their Employer's Liability insurance policy. Insurers will want details of the accident and will send their own agent to investigate the incident. The agent will be acting on behalf of the employer and should be given full co-operation. From this investigation the insurers will assess the likely damages and make suitable financial provisions and also use the information to decide whether to negotiate a settlement or to resist the claim if it is made.

7.3.3 In-house reporting

This has three functions:

1 It lets the senior managers know an accident has occurred and the facts of its circumstances.
2 It may indicate a shortcoming in safety arrangements within the organization and enable these to be corrected.
3 It provides the basic data from which domestic accident performance may be calculated.

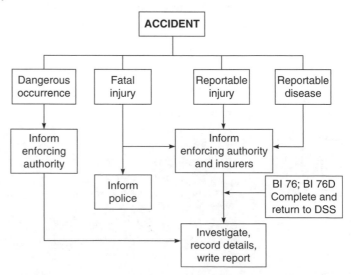

Figure 7.3 Diagram of action to be taken on learning of an accident

Part 3 Occupational health

Good health is an essential part of enjoying a good quality of life at home and at work. It is also an important factor in the viability of an organization. This fact is recognized by the Health and Safety Executive in their initiative titled 'Good Health is Good Business'.

Many work situations, by virtue of the layout of the working station or of the materials used, present a higher than normal risk to health. By understanding the characteristic of the materials in use and the likely reactions of the body the risks to health can be reduced to a minimum. In understanding the body's reaction to chemicals it is necessary to understand how the body functions.

Part 3 deals with these aspects and with the on-site actions that can be taken should ill-health or an injury be suffered.

The body

Appreciating how the body works, how it reacts to a range of substances used at work and being aware of the ways in which those substances can enter the body are important aspects of reducing to a minimum the causes of ill-health to which employees may be exposed.

This chapter looks at those particular aspects of ensuring good health at work.

**8.1
Functions of the
body**

The body is a complex organism comprising a great number of organs contained within a rigid structure (the skeleton) and held in place by various muscles. The different organs are all interdependent and play a specific role in the effective functioning of the body as a whole. However, the effectiveness of any of these organs can be adversely affected by conditions and substances met at work (and at home). The functions of some of the main organs and how they can be adversely affected are described below.

Organ	Function	Vulnerability
Bones	linked together they form the skeleton	they are brittle and can be broken by impact (blows) or sometimes by a muscular spasm
	red blood corpuscles are created in the bone marrow	this process is interfered with by toxic chemicals such as benzene and carbon monoxide and radioactivity
Skin	the protective layer that covers the outer surface of the body	can be penetrated by sharp objects and severe physical impact
		its protective fats can be dissolved by solvents resulting in dermatitis vulnerable to radiations
Muscles and tendons	control the movement of the various limbs	can be damaged by: – excessive loads (strains) – sudden jerks (sprains) – repetitive action (RSI and WRULD)
The gut	the digestive organs	means of changing ingested matter into substances needed by the body systems can be

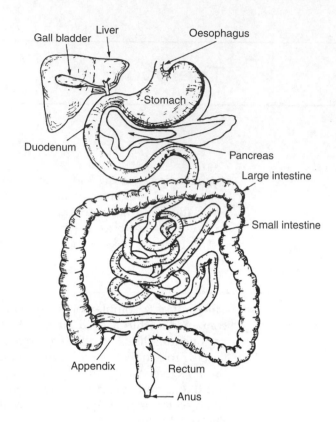

Figure 8.1 Diagram of digestive system

		damaged by ingested corrosive and poisonous substances
Liver	breaks down proteins from the gut, detoxifies body poisons and removes old red blood corpuscles	damaged by poisons such as organic solvents, certain metals, VCM and excessive alcohol
Kidneys	separates water and urea from body fluids and disposes of them	damaged by halogenated solvents and some heavy metals
Bladder	storage vessel for waste body fluids	prone to cancer from 2-naphthylamine
Lungs	the organ that takes oxygen from the air and transfers it to the blood	vulnerable to any respirable and inhalable fumes and dusts, particularly: – cancer from asbestos, radon and nickel – fibrosis from coal and silica dusts
Nerves	sensing organs that transmit messages to the brain	functioning impaired by exposure to toxic substances

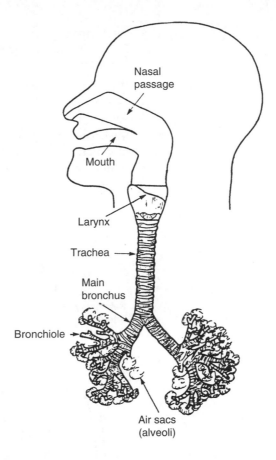

Figure 8.2 Respiratory system

Brain	the control centre for the whole body	narcotic effects of chlorinated solvents
		damaged by certain metals, carbon disulphide and carbon monoxide
Eye	the sight organ, delicate and exposed	vulnerable to: – dusts – flying particles – corrosive chemicals

Figure 8.3 Diagram of section through eyeball

Ear	the hearing organ incorporating the organ for balance	hearing acuity can be damaged permanently by prolonged exposure to high noise levels
Nose	the organ of smell	very sensitive olfactory nerves desensitized by hydrogen sulphide
Heart	the body's pump supplies blood and oxygen to the brain, various organs and muscles	its muscles can be affected by electric shock, resulting in acceleration or stopping (fibrillation) of its pumping action.

Figure 8.4 Diagram showing circulation of blood (RA — right auricle; LA — left auricle; RV — right ventricle; LV — left ventricle)

Each of these organs plays a unique role in the effective functioning of the body and must be protected from the agents that damage it. Some of those agents and the protective techniques are dealt with in the following sections.

8.2 Routes of entry

Hazardous and dangerous substances are able to enter the body by a number of different routes:

- Ingestion through the mouth and into the gut.
- Inhalation into the lungs.
- Absorption through the skin.
- Direct entry through open cuts and wounds.

Simple precautions can prevent these entries:

- Ingestion
  - no eating in the work place
  - personal hygiene, washing hands before eating
  - not smoking
- Inhalation
  - wear respiratory protection suitable for the particular substance
  - exhaust ventilation
  - fume and dust extraction

Target organs

- Absorption
  - wear protective gloves
  - wash all contaminated areas in soapy water
  - use a barrier cream
- Direct entry
  - get all cuts and wounds treated
  - keep all cuts and wounds covered when at work.

**8.3
Target organs**

Hazardous and dangerous substances do not attack the body indiscriminately. Different substances affect different organs although any particular substance may attack more than one organ. The hazards from substances can, broadly, be divided into seven groups and these are shown in the table below.

Type of hazard	Target organ	Reaction/symptom
Toxic	kidney, liver, bone marrow	attacks and affects the functioning of the organ
Carcinogenic	lung, liver, bladder	warts, ulcers, malignant growths
Corrosive	skin, lungs, stomach	destroys living tissues
Dermatitic	skin	inflammation of skin (dermatitis)
Irritant	skin, eyes, lungs	inflammation, dermatitis, fibrosis of lungs
Flammable	skin, whole body	burns
Radioactive	skin, sensitive organs such as bone marrow, eyes, gonads, etc.	leukaemia, cataracts, loss of fertility

Where a substance is used that affects more than one organ, it may be necessary to take separate precautions to protect each susceptible organ.

The hazard is related to the chemical characteristics of a particular substance.

Hazardous substances may occur in any of a variety of forms including solid, dust (particulate), gas, fume, liquid or a vapour (see Sections 11.1 to 11.7).

Many substances in a variety of forms can have adverse effects on the health of employees. Having an understanding of the substances and the forms that present hazards to health can take you half-way in eliminating or reducing the risks.

The effects of some of the more common forms of substances are considered in this chapter.

9.1
Causes of health hazards

Where, at work, an employee is subjected to certain conditions or materials there could be a reaction resulting in ill-health. In many of these cases the hazards may be known and the symptoms recognized. This section considers some of the main causes of occupational ill health.

Condition/material	**Body reaction**
Dusts	• when inhaled, affects the lungs causing pneumoconiosis. Particular dusts give rise to specific diseases:

        – asbestos   →   asbestosis, mesothelioma
        – silica   →   silicosis
        – coal dust   →   pneumoconiosis

Poisons
• ingested poisons:
   – can affect any one of a number of organs
   – the body absorbs a certain amount of the poison very quickly
   – the rest passes through and is evacuated
   – vomiting should *not* be induced since it can cause more damage than the original poison

Solvents
• enter the body by:
   – ingestion of liquids
   – inhalation of fumes
   – absorption through the skin
• can have:
   – narcotic effects on the nervous system
   – toxic effects on organs such as the liver, kidney or bone marrow
   – irritant effect by de-fatting the skin

Condition/material	Body reaction
Corrosives	• acids and alkalis • destroy living tissues • dilute by deluging with water • alkalis tend to be more persistent than acids • in both cases medical assistance should be sought
Irritants	• in the form of dusts or liquids may react with the skin and cause dermatitis • dusts when inhaled can cause irritation and fibrosis of the lungs

Carcinogens	• cause cancer in various organs including:
	  – the lungs through inhaling asbestos fibres
	  – the liver (as angiosarcoma) from VCM
	  – the skin from tar pitch
	  – the scrotum from mineral oils where there is poor personal hygiene
	  – the bladder by 2-naphthylamine

Gassing	• by the poisonous nature of the gas or fumes inhaled, i.e. chlorine, carbon monoxide, hydrogen sulphide, etc.
Asphyxia	• oxygen deficiency particularly due to heavier-than-air gases such as methane and carbon dioxide
Sensitization	• individuals can become sensitized to substances such as isocyanates, wood and flour dusts, moulds from rotting hay, colophony fumes from soldering, etc. • after sensitization, exposure to even the minutest concentration will cause a reaction • can seriously affect future working prospects especially where that particular substance is used

Bronchial asthma	• a prescribed industrial disease
	• caused by a range of materials including:
	  – isocyanates
	  – colophony fumes
	  – epoxy-resin curing agents
	  – wood and flour dusts
	  – animals and insects in laboratories

Diseases from metals	• diseases of varying seriousness can be caused by exposure to metals such as:
	  – lead
	  – mercury
	  – chromium

Condition/material	Body reaction

    – arsenic
    – manganese
    – nickel
    – cadmium
    – vanadium

Ionizing radiations
- emitted by radioactive materials
- ill-health effects range from:
  - damage to sperms and white blood cells
  - to nausea, vomiting, coma and death
- strict controls must be exercised where radio-active substances are used (see Section 9.2)

Vibrating tools
- cause a variety of injuries to hands and arms – known generically as 'hand-arm vibration syndrome' (HAVS)
- can cause constriction of the blood vessels in the hands, leading to a condition known as **vibration white finger** where the fingers go white and lose feeling
- operators with this condition should be moved to other work

Noise
- major effect is **noise induced hearing loss** (see Section 12.1)
- excessive noise can cause loss of concentration, fatigue and disorientation

Heat and humidity
- work at high temperatures and high humidity can result in:
  - cramps
  - heat stroke
  - exhaustion
- no standards to work to but the chill effect of moving air can help

Micro-organisms
- include a number of organisms:
  - viruses   →   hepatitis A and B, AIDS
  - bacteria   →   anthrax, legionella, leptospirosis, tetanus, etc.
  - fungi   →   Farmer's lung, ringworm
  - protozoa   →   malaria
  - nematodes   →   hookworm

Repetitive actions
- repeated forceful actions of upper limbs can cause:
  - tenosynovitis
  - carpal tunnel syndrome
  - writer's cramp
- sometimes referred to as repetitive strain injuries (RSI) or work related upper limb disorders (WRULD)

Condition/material	Body reaction
	• in early stages rest and removal to other work can result in recovery
Stress	• psychological reaction to factors often outside the individual's control such as: – demands above or below ability – the working environment – relations with fellow employees or the organization.

Where the symptoms persist after removal from exposure, seek medical advice.

Although there are many causes of health hazards at work, there are also a wide range of equipment and techniques to protect employees from them. Protection from these hazards is required by COSHH Regulations and is dealt with in Sections 11.1 to 11.8.

A nasty case of WRULD

9.2 Ionizing radiations

Radiations from radioactive decay cause the material they pass through to become ionized, hence the term 'ionizing radiations'. The effect on body tissues depends on:

• the nature or type of radiation
• the dose and duration of exposure
• whether the source is internal or external to the body.

9.2.1
Types of radiation

- **α (alpha)** – a particle whose radiation is stopped by a few centimetres of air, a sheet of paper or the outer layer of skin. However, ingested α particles can cause intense local radiation and immense damage to the affected tissues.
- **β (beta)** – have greater penetrating power than an α particle but the ionization caused is less severe.
- **γ (gamma)** – electromagnetic radiation with great penetrating power. Caused by radioactive decay and emits radiation all the time.
- **X-rays** – electromagnetic radiations whose penetrating powers depend on its energy. Commonly created in X-ray machines, the radiations cease when the machine is switched off.
- **Neutrons** – neutrons are emitted during nuclear fission and have very great penetrating powers. They can cause intense ionization.
- **Bremsstrahlung** – electromagnetic radiations produced by the slowing down of a β particle. They can have considerable penetrating powers.

9.2.2
Effects on the body

These are determined by the dose received, i.e. the type and intensity of the radiations and the period of exposure. Special instruments – ionization chambers or Gieger-Müller tubes – are used to measure the dose rate. Exposure levels can be determined by the use of *film badges*.

- Small localized exposure can cause:
  - redness of the skin
  - cataracts in the eyes
  - loss of fertility.

- General whole body exposure can result in:
  - nausea, vomiting and diarrhoea
  - cancer of the skin and other organs
  - leukaemia.

9.2.3
Uses

- Low power sources (α and β), usually encapsulated or sealed, are used for:
  - thickness gauges
  - smoke alarms
  - static-electricity eliminators.

- High power sources (γ, X-rays and neutrons) are used for:
  - medical diagnostic investigations
  - non-destructive testing of materials
  - high-technology production processes.

9.2.4
Precautions

- Low power sources:
  - follow suppliers instructions
  - appoint a Radiation Protection Supervisor (an employee who has been suitably trained) to oversee the safe use of the sources
  - when not in use, keep the sources in a secure store.

- High power sources:
  - appoint a Radiation Protection Adviser who must be suitably qualified and experienced. This expertise can be bought in from the Radiation Protection Division of Health Protection Agency (HPA-RPD) who provide this service as do a number of specialist consultancies
  - when not in use ensure the source is kept in a suitable shielded container in a secure store
  - when in use ensure that all employees except radiation specialists are clear of any area likely to be exposed to radiations
  - use sources only within properly shielded areas.

**9.2.5
Legislation**

The Ionising Radiations Regulations 1999 lay down the protective measures to be taken. These include:

- Restriction of employee's exposure.
- Control of access to areas where radiations may be present.
- Appointment of suitably trained or qualified persons to ensure safe use of sources.
- Implementation of rules for the safe use of sources.
- Training and instructing any employee who uses radiation sources.
- Measuring the exposure levels of employees working with radiations.
- Provision of medical examinations.
- Keeping accurate records of the use and locations of all sources.
- Reporting to the HSE any damage to or loss of a source.
- Investigating cases of over-exposure and taking corrective action.

Guidance on these Regulations is given in HSE publication L121, 'Work with ionizing radiation'.

Where the use of radiation sources is contemplated, advice should be sought from the supplier, HPA-RPD or suitably qualified consultant.

**9.3
Hazards from
non-ionizing
radiations**

Non-ionizing radiations have a longer wavelength than ionizing radiations and do not cause ionization. However, they can give rise to serious risks to health and suitable precautions should be taken when they are used. They occur both naturally and in industry where they have a number of important uses.

Table 9.1 summarizes the types of non-ionizing radiation, their uses, hazards and precautions to be taken.

Table 9.1 Summary of non-ionizing radiations, uses and precautions

Type of radiation	Occurrence	Uses	Hazards	Precautions
Ultraviolet	Solar radiations Ultraviolet emitting equipment Welding arc flash	Sun beds Curing inks and paints	Erythema (sun burn) Arc eye Cataracts Skin cancer	Keep exposed parts covered Wear protective goggles Wear protective glasses Use barrier cream Keep equipment that generates ultra-violet fully enclosed with interlocks
Visible light	Solar radiations Lighting equipment	Illumination	Possible retinal damage from badly positioned luminaries	Control intensity of lights and positioning of luminaries
Infra-red	Any high temperature surfaces Special infra-red generators	Process heating Heat-seeking devices Security and safety devices	Burns Cataracts	Provide guards Use insulated gloves
Radio frequency	Radio transmitters Radar Dielectric heaters Broadcasting equipment Microwave heaters	Communications Navigation systems PVC welding Cooking	Whole and partial body heating RF burns Cataracts	Provide screens Earth metal parts Check for radiation leakages
Electro-magnetic	Overhead power lines Induction heaters Zonal smelters Display screen equipment	Power transmission Domestic appliances	Nervous system dysfunction Irritation of facial skin	Move away Earthing appliances Very low risk, try earthing screen
Lasers	Special generating equipment	Illuminated displays Lining up plant and buildings Cutting wood and metals Entertainment Pointers	Mono-chromate sources over very wide range of wavelengths Delivers intense power to very small area Can cause corneal and retinal damage and skin burns	Enclosure of equipment, safety interlocks and shutters, wear goggles and gloves Competent trained operators Display warning signs

10 Health protection

There are a number of well established techniques for looking after the health of employees. These include taking action to prevent ill-health, providing means to prevent employees coming into contact with hazardous substances and ensuring that if they are hurt in any way their injuries are properly treated.

This chapter looks at the means to protect employees' health.

10.1 First aid

First aid is defined as:

- emergency treatment until the arrival of medical practitioner or nurse
- treatment of minor injuries that do not warrant medical attention or would otherwise receive no treatment.

First aiders trained by the Red Cross or St. John's Ambulance Brigade are no longer recognized unless their trainer had been approved by the HSE.

All employers with 10 or more employees must have available a copy of the 'new' accident book, form BI 510, in which an employee can record the fact of his/her injury. Copies of BI 510 are obtainable from HSE Books.

The first-aid facilities to be provided are outlined in the *Health and Safety (First Aid) Regulations 1981*, with greater detail being given in an Approved Code of Practice and Guidance 'First aid at work', HSE publication L74. The recommendations include:

- The extent of the facilities should depend on the risks faced, i.e. the higher the risk, the more extensive the facilities.
- There should be an adequate number of suitable persons to render first aid. Suitable persons should be trained by HSE-approved trainers and qualified to HSE's standards for first aiders – one should be appointed per fifty employees for low-risk work, with the number increasing as the risks increase.
- There should be a first-aid room if:
  - site is high-risk
  - site is remote from a hospital, e.g. in country areas
  - access to a hospital or doctor is difficult, e.g. in areas with bad traffic congestion
  - the number employed on the site warrant it.

... a suitable person ...

- Where employees work away from base:
  - if work area is low-risk – no facilities
  - if work area is on another employer's premises – use local facilities
  - if area of work is high-risk or does not have access to first-aid facilities, a first-aid kit to be carried. (For contents see below.)
- First-aid boxes must:
  - be constructed to protect contents
  - be kept replenished
  - contain a first-aid guidance card
  - not be used for keeping anything other than first-aid items.
- Where more than one employer occupies a building or site, common shared facilities can be provided.
- Employees should be informed of first-aid facilities and their location.
- First-aid facilities may be made available to visitors, contractors, etc. on premises.
- Where a first-aid room is provided it must:
  - be under the control of a first aider or nurse
  - have a first aider available at all times anyone is working on the premises
  - where the first aider is temporarily absent, have a responsible *appointed person* to deal with any first-aid treatment needed
  - be conveniently situated for ambulance access
  - be large enough to accommodate a couch
  - have doorways wide enough to allow passage of a wheelchair
  - be decorated with surfaces that can be cleaned easily
  - have hot and cold water for washing
  - be clearly identified
  - give location of nearest first aider
  - be provided with a treatment book which can be the employer's own day book to record treatments.
- First aiders must:
  - be trained on an HSE-approved course
  - have received specific training where special hazards arise

– record all treatments given
– receive regular refresher training
– Advice is given in HSE publication HSG212 *The training of first aid at work.*

10.1.1
First-aid boxes

First-aid boxes should, as a minimum, contain:

- a guidance card
- 20 individually wrapped sterile adhesive dressings
- 4 individually wrapped triangular bandages
- 6 safety pins
- 6 medium-sized individually wrapped sterile unmedicated wound dressings
- 2 large individually wrapped sterile unmedicated wound dressings
- 3 extra-large individually wrapped sterile unmedicated wound dressings
- 1 pair of disposable gloves
- 2 sterile eye pads.

A supply of tap or sterile bottled water for irrigating eyes should be available. Additional items that can be provided include:

- a stretcher or other means of carrying a patient
- a pair of blunt-nosed stainless steel scissors
- a disposable plastic apron and gloves
- blankets
- a suitable bin for disposal of used swabs and dressings.

Travelling first-aid kits to contain:

- a first-aid guidance card
- 6 individually wrapped sterile adhesive dressings
- 1 medium sterile unmedicated dressing
- 2 rolls of bandage 1" wide
- 2 safety pins
- individually wrapped moist cleansing wipes
- 1 pair of disposable gloves.

Employers are given a certain amount of discretion to match the facilities they provide with the risks faced. First-aid box suppliers can be helpful but watch the pushy salesman.

10.2
Personal
protective
equipment

General requirements to provide personal protective equipment (PPE) are contained in the *Personal Protective Equipment at Work Regulations 1992.* However, specific regulations dealing with particular hazards take precedence over those general requirements, namely:

The Control of Lead at Work Regulations 2002
The Ionising Radiations Regulations 1999

The Control of Asbestos at Work Regulations 2002
The Control of Noise at Work Regulations 2005
The Construction (Head Protection) Regulations 1989
The Work at Heights Regulations 2005.

These latter Regulations are considered separately elsewhere and are not covered in this section.

In providing protection against a hazard, an employer's first priority must be to protect the workforce as a whole rather than individuals. The use of PPE should only be considered if more global methods of protection are not reasonably practicable or feasible.

With all PPE, the supplier will advise on the most appropriate type to provide the protection needed, and may offer a choice of material, design, colour, etc. However, there are some general principles that should be followed.

To be effective PPE must:

- be appropriate for the hazard faced
- be of a material that will resist that hazard
- be suitable for the person using it
- not interfere with the operator performing his/her functions
- be of robust construction
- not interfere with other PPE being worn at the same time
- not increase the risks to the wearer.

... appropriate for the hazard faced ...

PPE should be:

- provided free of charge
- personal issue or given suitable hygiene treatment between uses
- used only for its intended purposes
- kept in good repair
- repaired or replaced if damaged
- kept in suitable accommodation when not in use.

Operators who use PPE should be:

- informed of the hazards faced
- instructed in the preventative measures taken
- trained in the proper use of the equipment
- consulted and allowed a choice of PPE subject to its suitability
- taught how to maintain the equipment and keep it in good order
- instructed to report any deterioration or damage.

Examples of the protection provided by a range of types of PPE are:

Part of the body	Hazard	PPE
Head	falling objects	hard hats
	confined spaces	bump caps
	hair entanglement	caps, hair net, hair cut
Hearing	excessive noise	ear muffs, ear plugs
Eyes	dust and grit, flying particles	goggles, face shields
	radiations, lasers, arc welding	special goggles
Lungs	dusts	face masks, respirators
	fumes	respirator with absorbent filter (limited effectiveness)
	toxic gases and oxygen-deficient atmosphere	breathing apparatus
Hand	sharp edges and burrs	protective gloves
	corrosive chemicals	resistant gloves
	low/high temperatures	insulating gloves
Feet	slipping, sharp items on the floor, falling objects	safety shoes
	liquid-metal splashes	gaiters and safety shoes
Skin	dirt and mild corrosives	barrier creams
	strong corrosives and solvents	impermeable barriers such as gloves and apron

Trunk and body	solvents, moisture, etc.	aprons, overalls
Whole body	hostile atmosphere (toxic fumes/radioactive dusts)	pressurized suits
	falls	safety harness
	moving vehicles	high-visibility clothing
	chain saws	special protective clothing
	high/low temperatures	heat-resistant/warm clothing
	inclement weather	all-weather clothing

The following publications give advice on PPE:

- HSE's booklet L25 'Personal Protective Equipment at Work'
- HSE's booklet L101, 'Safe working in confined spaces'
- HSE's booklet L102, 'A guide to the Construction (Head Protection) Regulations 1989'
- HSE's booklet HSG 53 'The selection, use and maintenance of respiratory protective equipment'.

10.3
Safe use of display screen equipment

The increasing use at work of computers, word processors and other electronic graphic display equipment has brought in its wake a range of health hazards. Requirements to mitigate the effects of these hazards are contained in the *Health and Safety (Display Screen Equipment) Regulations 1992* as amended. These Regulations are supported by an HSE guidance booklet no: L26 'Display screen equipment work'.

The Regulations:

r.1 Define

- 'display screen equipment' (DSE) as any alpha-numeric or graphic-display screen regardless of the use or process, i.e. not only word processors but computer-design and computer-controlled machinery
- 'operator' as a self-employed user
- 'user' as any person who uses DSE for a significant part of their normal work
- 'work station' as any equipment, or part of it, used in connection with work on a DSE.

r.2 Requires the carrying out of a risk assessment and the reduction of identified risks so far as is reasonably practicable.

r.3 Requires the taking of precautions which are enlarged on in the Schedule to the regulations and include:

- Display screen
  - to swivel and tilt
  - characters well defined and clearly formed
  - no flicker
  - adjustable contrast
  - free from reflections and glare
  - kept clean and free from dust
- Keyboard
  - tiltable
  - independent of the screen
  - space to support hands and arms
  - matt surface
  - key symbols clearly readable
- Work desk
  - large enough to accommodate documents and adjustments of equipment
- Work chair
  - stable
  - adjustable height
  - adjustable back rest
  - footrest
- Environment
  - space for the operator to change position
  - good lighting without background reflections or glare
  - noise emitted not to distract attention
  - not emit heat to cause the user discomfort
  - radiations – other than visible – reduced to negligible level
  - maintain humidity at a comfortable level
- User interface
  - software suitable for the task
  - compatible with operator's level of knowledge/experience
  - format and pace of data presentation adjustable to suit operator.

r.4 Time

- limit time on continuous keyboard work and ensure breaks at regular intervals. Time between breaks determined by the type of work but should not exceed one hour.

r.5 Visual

- check that the operator has normal eyesight. Indications that this is not so are eye strain, headaches, blurring of the vision
- if not, eyesight tests to be carried out by competent person and the provision of special vision-corrected glasses (employer

bears the cost of standard frames but employee contributes extra if they insist on special frames)
- background lighting should be general without points of light that could cause glare on the screen
- the screen should be positioned or adjusted to eliminate glare or reflections from lights or windows
- operators should be able to adjust screen contrast to suit themselves
- screen should be kept clean and free from dust build-up.

r.6 Training

- The user is to be provided with adequate training before and during employment and after changes to the work situation.

r.7 Information

- hazards associated with the workstation
- precautions provided to avoid hazards.

Considerations given should include:

Equipment

- screen should give a stable picture. If instability, flicker, jump, jitter, etc. persist seek advice from supplier
- polarity should be adjustable to suit the user
    positive polarity = dark characters on light ground
    negative polarity = light characters on darkground
- keyboards should suit the user or the user be trained to the particular keyboard provided
- work desks need to have adequate surface space to accommodate the keyboard, working materials, document-holder, etc.
- surface of work desks should be plain, smooth and easy to clean
- the workstation should be spacious with clear space beneath to allow freedom of leg and body movement.

Environment

- noise should be kept to a minimum. Adjacent noisy equipment (printers, etc.) should be enclosed in soundproof cabinet
- sound-absorbing screens help reduce noise from other equipment
- electronic equipment can make the atmosphere dry; ventilation or other means (plants) should be used to maintain humidity
- radiations – electromagnetic, ionizing and radio frequency – emitted by DSE are well below national recommended levels and less than everyday background levels
- static electricity build-up can cause skin irritation in some people. Reducing the build-up on the screen by wiping it with a damp cloth can help.

User interface

- users must be trained in the particular software
- software must be suitable for the task
- software must be capable of being used at a rate to suit the user
- there should be effective back-up to resolve problems that the user cannot deal with.

Part 4   Safety technology

The need for legislation to give work people protection in the jobs they were forced by circumstances to undertake became a matter of increasing urgency as the Industrial Revolution got under way and brought in its train an enormous toll of horrendous injuries. Machines were being developed to increase the output per worker that paid scant regard to the worker's health or safety.

Initially the major area of concern was the dangerous nature of the machinery but as we became more industrialized and the machinery more complex, a vast range of chemicals has been introduced to aid productivity and to improve the finished product-chemicals whose short and long term health effects are often little understood.

In modern industry and commerce many of the methods of protection have a technical background and a considerable degree of expertise is necessary to ensure those methods are effective against the hazards faced.

This part deals with various of those hazards and the techniques needed to give an acceptable level of protection.

11 Chemicals

Of the vast range of chemicals and substances in use in industry, commerce and other areas of employment, many are harmless. However, there are also a considerable number of substances in use which have a high potential to cause harm and ill-health – 'substances hazardous to health' usually referred to as *hazardous substances*.

There is a considerable range of legislation dealing with chemicals which has as its objective the protection of those who handle these substances by ensuring that employers who make or use these dangerous substances have in place precautions and systems of work that reduce to a minimum the chances of the substances harming anybody.

Excluded are the substances that can cause physical damage, such as by explosions, i.e. flammable gases and explosives, which have separate laws.

It is convenient to consider the problems of the safe use and handling of hazardous substances in three parts which conveniently coincide with the relevant laws:

1   Safe use of substances
    – *The Control of Substances Hazardous to Health Regulations 2002 (COSHH)* – see Section 11.1
2   Supply of substances (covering packaging and labelling)
    – *The Chemicals (Hazard Information and Packaging for Supply) (Amendment) Regulations 2002 (CHIP 2002)* – see Section 11.2
3   Transport of substances by road and rail
    – *The Carriage of Dangerous Goods (Classification, Packaging and Labelling) and Use of Transportable Pressure Receptacles Regulations 1996* – see Section 11.3

Where specific problems arise in the use, handling or transporting of hazardous substances reference should be made to the particular relevant Regulations. Copies of the Regulations and supporting advisory Codes of Practice and Guidance should be held by the local reference library. However if they are not available there, copies can be purchased from TSO and/or HSE Books.

**11.1
Safe use of
chemicals**

The main legislation concerned with safety in the use of chemicals is *The Control of Substances Hazardous to Health Regulations 2002* (COSHH) which is aimed at reducing the likelihood of ill-health that can result from lax ways of handling and using some of the nastier chemical substances. The main requirements of these Regulations are summarized below.

r.2   Defines, inter alia:

- a substance hazardous to health as:
  - a substance listed in part 1 of the Approved Supply List as dangerous for supply within the meaning of CHIP 2002 – see Section 11.5
  - a substance given a workplace exposure limit (WEL)– see Section 11.6
  - a biological agent
  - any dust in substantial concentrations (roughly more than 10 mg/cu m of air which is a thick cloud)
  - any other substance causing a similar hazard to health
- a biological agent as:
  - any micro-organism, cell culture or human endoparasite including genetically modifed organisms that may create a risk to human health
- a carcinogen as:
  - any substance likely to cause cancer and classified in r4(4) or listed in Schedule 1 of CHIP 2002 as carcinogenic
- workplace exposure limit as the exposure limit specified in HSE's publication EH40:2005 as updated.

r.3   Places duties on the employer to ensure that chemicals are used and handled in such a way that no one, whether an employee or not, is put at risk.

r.4   Bans:

- the importation (except from an EU country) of:
  - 2-naphthylamine ⎱ their salts and any compound
  - benzidine ⎰ containing more than 0.1% of
  - 4-aminodiphenyl ⎰ them
  - 4-nitrodiphenyl ⎰
  - matches made with white phosphorus
- the use in certain processes of:
  - the above substances
  - substances with free silica
  - carbon disulphide
  - mineral oils for use on the spindles of self-acting mules
  - ground or powdered flint or quartz
  - white phosphorus
  - hydrogen cyanide
  - benzene and substances containing it
  - a range of solvents for use in surface cleaning and the cleaning of fabrics.

r.5 Excludes:

- lead which is subject to the *Control of Lead at Work Regulations 2002*
- asbestos subject to the *Control of Asbestos Regulations 2006*
- siliceous dusts subject to the *Coal Mines (Respirable Dusts) Regulations 1975*
- substances whose sole risk is physical, i.e. radioactive, explosive, flammable, low or high temperature
- substances used for medical treatment
- substances used in mines under mines legislation.

r.6 Requires employers:

- to carry out a suitable and sufficient risk assessment wherever employees may be subject to risk to health from hazardous substances.

The risk assessment should consider, inter alia:

- the hazardous properties of the substance
- safety data provided by the supplier
- extent of exposure
- method of use and how much
- high hazard risk activities such as cleaning and maintenance
- relevant WELs (or EU equivalent)
- the effect of preventative and control measures
- the results of any health surveillance
- the results of any exposure monitoring of the area or activity
- the combined hazardous effects where exposure is to a number of substances
- for biological agents, their approved classification.

The risk assessment to be reviewed:

- regularly
- if existing assessment no longer valid
- if there has been a change in the work, materials or methods
- monitoring results show a need

and changes made as necessary.
Where there are 5 or more employees the risk assessment must be recorded together with the action taken. See also Section 3.4.

r.7 Requires employers to prevent or control, in the following order of priorities, the exposure of employees to substances hazardous to health by:

- using non-hazardous or less hazardous materials
- the design and use of less hazardous equipment

- controlling exposure at source
- adjusting working methods and systems
- the use of PPE.

Techniques for protecting employees from exposure to include:

- arrangements for the safe handling, storage and transport of hazardous substances and wastes containing them
- implementing safe maintenance procedures
- reducing:
  - number exposed
  - level of exposure
  - duration of exposure
  - quantity of hazardous substance in use
- controlling the working environment by providing ventilation
- instituting high hygiene standards including washing facilities.

Where there is a carcinogen, take the following precautions:

(a) total enclosure of the process and handling system
(b) prohibit eating, drinking and smoking in the area
(c) regular cleaning of work area floors, walls and other surfaces
(d) identify areas at risk with suitable warning notices
(e) ensure high standards of safety in storage, handling and disposal of substances.

Where biological agents are used or are present, exposure should be prevented. If this is not possible the employer must take further precautions to:

- identify the area and post warning signs
- specify decontamination and disinfection procedures
- arrange for the safe collection, storage and dispatch of contaminated waste
- check for the presence of biological agents in adjacent areas
- lay down procedures for transporting and using biological agents
- have available suitable vaccines for exposed employees
- implement high hygiene standards including:
  - adequate washing and toilet facilities
- prohibiting eating, smoking, drinking and the use of cosmetics in the risk area
- take the action specified in Schedule 3 of the Regulations where humans or animals are infected.

Where a substance hazardous to health is used:

- the principles of good practice set out in Schedule 2A (and cited below) must be followed

- if a WEL is specified, that exposure must not be exceeded
- where the hazardous substance is allocated a risk phrase R42, R42/R43, R45, R46 or R49 or is an 'asthmagen' exposure must be reduced to as low a level as is reasonably practicable.

Any PPE provided must:

- be suitable for its purpose
- comply with the PPE Regulations (see Section 10.2)
- provide suitable protection against the substance concerned
- if respiratory protection equipment, it must be of a type approved by the HSE.

r.8 Employers must ensure that all control measures and equipment provided are properly used.
Employees must use protective equipment properly, return it to its store and report any defects or damage found.

r.9 All equipment and plant provided in compliance must be kept in an efficient state, effective working order, good repair and a clean condition. Where the precautions rely on systems of work or supervision, they must be kept under review and revised as necessary.

Engineering controls must be examined and tested as follows:

- local exhaust ventilation (LEV)   – every 14 months
- ventilation systems for:
  - blast cleaning of metal castings   – every month
  - grinding and polishing of metals   – every 6 months
  - dust and fume from non-ferrous castings – every 6 months
  - jute cloth manufacture   – every month

Ah! a cheeky little mix

... regularly monitored ...

RPE must be examined and tested at suitable intervals.
A record must be kept of the examinations and tests and held for five years.
Where PPE is provided it must be:

- properly stored
- checked at suitable intervals
- have any defects repaired or be replaced
- if contaminated, kept separate from uncontaminated equipment, decontaminated and cleaned or destroyed as appropriate.

r.10 Where the risk assessment shows that protection is necessary for employees who are exposed to hazardous substances, the atmosphere should be monitored and readings recorded:

- at regular intervals
- when changes occur that affect exposure
- in processes involving;
  - VCM – continuously
  - spray from electrolytic chromium processes – every 14 days.

Principles of good practice for the control of exposure to substances hazardous to health

- Design and operate processes and activities to minimize emission, release and spread of substances hazardous to health.
- Take into account all relevant routes of exposure – inhalation, skin absorption and ingestion – when developing control measures.
- Control exposures by measures that are proportionate to the health risk.
- Choose the most effective and reliable control options which minimize the escape and spread of substances hazardous to health.
- Where adequate control of exposure cannot be achieved by other means, provide in combination with other control measures suitable personal protective equipment.
- Check and review regularly all elements of control measures for their continuing effectiveness.
- Inform and train all employees on the hazards and risks from the substances with which they work and the use of control measures developed to minimize the risks.
- Ensure that the introduction of control measures does not increase the overall risk to health and safety.

Monitoring can be carried out by:

- 'grab sampling' using a pump to draw air through a stain indicator tube; the length of the stain gives a measure of the concentration. However:
  - measure is not very accurate
  - stain tubes can only be used once

- stain tubes can only be used for one substance
- necessary to keep stocks of tubes for each substance monitored
- can give false reading with mixtures of substances
- passive or activated carbon collector used with air pump:
  - requires complex chemical analysis to obtain concentration
- electronic meters
  - expensive to purchase
  - give immediate and accurate reading in parts per million (ppm)
  - can be used for a range of substances at any one time
  - can be used continuously
  - may include facility to integrate daily exposure levels
  - contain an automatic alarm if pre-set limit exceeded.

Records of monitoring must be kept:

- for personal exposure records of identified employees
  - for forty years
- all other readings such as general area monitoring
  - for five years.

r.11 Health surveillance where appropriate and if listed in Schedule 6, should be carried out:

- by a doctor
- every twelve months where an employee works in specified processes with:
  - vinyl chloride monomer
  - nitro or amino derivatives of phenol and benzene
  - potassium or sodium chromate or dichromate
  - ortho-tolidine and its salts
  - dianisidine and its salts
  - dichlorbenzidine and its salts
  - auramine
  - magenta
  - carbon disulphide
  - disulphur dichloride
  - benzene and benzol
  - carbon tetrachloride
  - trichloroethylene
  - pitch
- where there is exposure to any other hazardous substance which:
  - can cause an identified disease or ill health condition
  - is likely, under working conditions, to result in ill health
- there are techniques for detecting the condition
- diagnosis does not put the employee at risk

- the examining doctor may:
  - determine the frequency of examinations
  - place conditions on the employee continuing with his employment when the employer must ensure those conditions are met
  - continue examinations after the exposure of the employee has ceased
- where an employee is exposed to a prescribed substance that employee shall provide the examining doctor with information about his general health
- if an employee is found to have an identified disease caused by exposure at work, the employer shall:
  - ensure the employee receives information and advice about the condition
  - review the risk assessment
  - ensure medical surveillance continues as long as the doctor recommends
  - consider giving the employee other work
  - arrange for other similarly exposed employees to have medical surveillance.
- records of health surveillance to be kept for forty years
- medical examinations should take place during normal working hours
- employees must be allowed to see their medical records
- medical adviser may inspect the work place if he/she wishes
- appeal against a medical decision must be made to the HSE.

r.12 Employees exposed to hazardous substances are to:

- be given such information, instruction and training so that they:
  - know the names of the substances
  - know the risks they present
  - are aware of WELs or other occupational exposure limits
  - know legislative requirements concerning the substances
  - have access to the findings of the risk assessment
  - know the precautions and action being taken to safeguard him
  - have access to the results of monitoring of exposure, especially if WELs have been exceeded
  - have access to collective information on the health surveillances being carried out
  - have written instructions when dealing with group 4 biological agents
  - are provided with updated information etc. when work changes occur.

Pipes and vessels containing substances hazardous to health to be clearly marked with the contents and their hazards. Others employed to do the same work must be given the same information instruction and training.

r.13 To deal with possible accidents, incidents and emergencies, the employer shall:

- have in place:
  - appropriate first aid facilities
  - relevant safety drills which must be tested regularly
- have available information on emergency arrangements covering:
  - work hazards
  - hazard identification arrangements
  - particular hazards that may arise during an accident, incident or emergency
- have established suitable warning and communication systems to assist in remedial action and rescue operations
- ensure this information is:
  - made available to the emergency services
  - displayed in the workplace
- take immediate steps to:
  - mitigate the effects of an emergency
  - restore the situation to normal
  - inform the affected employees
  - only allow suitably trained and equipped employees into the affected area until the situation returns to normal
- where a biological agent has been released to inform his employees of:
  - the cause of the incident
  - the measures taken to rectify the situation.

Where, except for carcinogenic and biological agents, the risk assessment shows that the quantity of each hazardous substance represents only a slight risk or the arrangements in place control the risk, the employer does not have to take mitigating action.

Employees are required to report the release of a biological agent that could cause severe human disease.

r.14 With certain exceptions, notice must be given to specified authorities of intention to fumigate using:

- hydrogen cyanide
- phosphine
- methyl bromide.

If fumigation is:

- in harbour area, notice to be given to:

- harbour authority
- HSE or local authority inspector and
- if of a sea-going ship:
  * chief fire officer of area
  * officer in charge of Customs and Excise
- if of a building:
  * chief fire officer of area
• elsewhere, notice to be given to:
  - local police
  - HSE inspector for the area

...give notice of
fumigation...

The information to be given includes:

• name, address and telephone number of the fumigator
• name of the client
• address and description of the premises to be fumigated
• date of the fumigation
• name of the fumigator operator
• the fumigant to be used.

Before fumigation starts, warning notices must be posted round the premises concerned.

r.21 In proceeding for an alleged infringement, a defence is allowed that to avoid causing the breach:

• all reasonable precautions were taken
• all due diligence was exercised
• proof must be provided by the accused.

The nub of these Regulations is the identification of hazards from substances that are hazardous to health so that precautionary procedures can be introduced. In addition to professional knowledge of the characteristics of the substances which may be obtained from the supplier's safety data sheets (SDSs), the main vehicle for identifying the hazards is the risk assessment (see Section 3.4).

11.2 Labelling of chemicals for supply and use

The *Chemicals (Hazard Information and Packaging for Supply) Regulations 2002* (CHIP 2002) are concerned with ensuring that hazardous chemicals when packaged for sale, both to employers and to consumers, are properly packaged and adequately labelled. Also that safety data on the chemicals are provided to the users.

The content of these Regulations is summarized below.

r.2   Defines a number of phrases used in the regulations including:

- *approved classification and labelling guide* as the HSC publication 'Approved Guide to the Classification and Labelling of Dangerous Substances and Dangerous Preparations' (5th Edition)
- *Approved Supply List* as the document entitled 'Information Approved for the Classification and Labelling of Dangerous Substances and Dangerous Preparations', Guidance on the Regulations (8th Edition) (see Section 11.5)
- *category of danger ... for a dangerous substance or dangerous preparation ...* as one of the following:
  - physico-chemical properties
    * explosive
    * oxidizing
    * extremely flammable
    * highly flammable
    * flammable
  - health effects
    * very toxic
    * toxic
    * harmful
    * corrosive
    * irritant
    * sensitizing
    * carcinogenic
    * mutagenic
    * toxic for reproduction
  - environmental effects
    * dangerous for the environment
- *Community workplace exposure limit*
  - the exposure limit for a substance established by the EU
- *EC number* as:
  - for substances on the approved list, the number quoted

  – for substances not on the approved list, the number given in the European Inventory of Existing Commercial Chemical Substances (EINECS)
  – for new substances, the number given in the European List of Notified Chemical Substances (ELINCS)
- *indication of danger* as shown by the symbols used to denote the different types of danger with an identifying letter. See Section 11.4
- *preparation* as a mixture or solution of two or more substances
- *risk phrase* as a phrase describing the risk posed by a substance and designated by the letter R with a number – the phrases and number are given in the Approved Supply List (see Section 11.5)
- *safety phrase* as a phrase describing the safety precautions to be taken and designated by the letter S with a number – the phrases and numbers are given in the Approved Supply List (see Section 11.5).
- *supply* means making a substance or preparation available to another and includes importation
- *package* includes the packaging in which the substance is supplied, including pallets, but not the wrapping provided by a retailer.

r.3 Specifies that the regulations apply to any dangerous substances or dangerous preparations for supply except, *inter alia*:

- those covered by other legislation such as:
  – radioactive substances
  – cosmetic products
  – medicines
  – controlled drugs
  – special waste
- samples taken for examination
- munitions
- fireworks.

r.4 Suppliers must not supply dangerous substances or dangerous preparations unless they have been classified in accordance with these or other regulations.

r.5 Suppliers of dangerous substances or dangerous preparations must provide customers, who intend using the substance at work, with SDSs containing information on the following:

- name of the company supplying the substance
- the substance's name
- its ingredients
- hazard identification
- first-aid measures
- fire-fighting measures

- accidental release measures
- handling and storage
- exposure controls and personal protection
- physical and chemical properties
- stability and reactivity
- toxicological effects
- ecological information
- disposal considerations
- transportation information
- regulatory information
- other information.

Safety data sheets must be free of charge, kept up to date and given to customers before the first delivery and be in English.

r.6   Advertisements of dangerous substances must mention the danger or hazard.

r.7   Packaging for dangerous substances must:

- be suitable for the substance contained in it
- prevent leakage or escape
- be capable of withstanding expected handling
- have re-usable closures that seal repeatedly
- not be of a material that will react with the contents.

r.8   Labels for dangerous substances must carry the following information:

- name, address and telephone number of supplier
- name of the substance or trade name of preparation
- indication of danger and appropriate hazard symbol
- risk phrases set out in full
- safety phrases set out in full
- EC number if any
- substance classified as:
  - carcinogenic
  - mutagenic   } must be labelled as *restricted to professional users*
  - toxic for reproduction
- labels for dangerous preparations must contain the following additional information:
  - if for sale to the general public, the nominal quantity.

r.9   Gives information required on labels with specific particulars for dangerous preparations that:

- are to be supplied to the public
- are to be applied by spraying

Hazardous
substances…properly
labelled

- have cumulative effects
- could cause harm to breast-fed babies
- are pesticides
- are paints and varnishes containing lead
- are cyanoacrylate based adhesives
- are preparations containing:
  - isocyanates
  - epoxy constituents
  - active chlorine
  - cadmium and alloys for soldering or brazing.
  - preparations not yet fully tested
  - sensitizers
  - halogenated hydrocarbons
  - cement and cement products
  - substances given the risk phrase R67.

r.10 Sets down details of the printing and attaching of labels.

r.11 Requires:

- child-resistant fastenings to comply with the appropriate harmonized or international standard
- packages containing a dangerous substance not be of a shape to attract children
- all packages containing dangerous substances to carry a tactile warning to BS EN ISO 11683 for the benefit of the partial sighted and the blind.

r.12 Requires the person who first supplies a dangerous preparation to keep the information used in:

- classifying the dangerous preparation
- its labelling
- child resistant enclosures
- tactile warning
- the preparation of a SDS

for at least 3 years after the date of the last supply of the preparation. A copy of this information must be readily available for inspection by the enforcing authorities.

r.15 Allows defence of *took all reasonable precautions and exercised all due diligence to avoid the commission of that offence.*

Schedule 1 of CHIP 2002 provides a list of the types of danger from chemicals met in the workplace. To each type of danger is given a symbol letter that can be used on labels and warning signs.

Category of danger	Symbol letter
Physio-chemical properties	
Explosive	E
Oxidizing	O
Extremely flammable	F+
Highly flammable	F
Flammable	None
Health effects	
Very toxic	T+
Toxic	T
Harmful	Xn
Corrosive	C
Irritant	Xi
Sensitizing by inhalation	Xn
Sensitizing by skin contact	Xi
Carcinogenic: categories 1 and 2	T
category 3	Xn
Mutagenic: categories 1 and 2	T
category 3	Xn
Toxic for reproduction: categories 1 and 2	T
category 3	Xn
Dangerous for the environment	N

Schedule 2 gives the symbol letters together with the equivalent warning symbol

Danger	Letter	Symbol
Explosive	E	
Oxidizing	O	
Extremely flammable	F+	
Highly flammable	F	
Very toxic	T+	
Toxic	T	
Harmful	Xn	
Corrosive	C	
Irritant	Xi	
Dangerous for the environment	N	

The skull and crossed bones symbol is also used with the following written abbreviated notes:

Note	Hazard	Description of hazard
Carc Cat 1	Category 1 carcinogen	Chemicals that may cause
Carc Cat 2	Category 2 carcinogen	cancer or increase its incidence
Muta Cat 1	Category 1 mutagen	Chemicals that induce
Muta Cat 2	Category 2 mutagen	heritable genetic defects or
Muta Cat 3	Category 3 mutagen	increase their incidence
Repr Cat 1	Category 1 reproductive toxins	Chemicals that produce or increase the incidence
Repr Cat 2	Category 2 reproductive toxins	of non-heritable effects in progeny and/or an impairment in reproductive functions or capacity

Also used is the cross symbol with a note:

Carc Cat 3   Category 3 carcinogen

Pesticides are not given separate designatory letters or symbols but must be classified as either very toxic, toxic or harmful, as determined by tests outlined in the Regulations.

The designatory letter and symbol must be displayed on the labels of the packaging of substances dangerous for supply.

11.3 REACH

On 1 June 2007, the EU brought into effect *Regulations concerning the Registration, Evaluation, Authorization and Restriction of Chemicals* (REACH). These have the aim of improving the protection of makers and users from the damaging effects of the chemicals with which they may come into contact. As an EU regulation, it has direct applicability in all Member states and takes precedence over local laws and arrangements. Many of REACH requirements are already contained in the UK legislation, so it is likely that only relatively minor amendments will be necessary to COSHH, CHIP, NONS, etc.

The basic requirement of REACH is that all chemical substances made, imported, used or stored in quantities greater than 1 tonne per year have to be registered. Those substances included in the European Inventory of Existing Chemical Substances (EINECS) are assumed to be registered. This inclusion should be checked since *no registration – no market*. Applications to register for non-registered chemicals should be made by December 2008

Chemical producers and suppliers are required to:

- submit a list of existing chemical substances quoting EINECS and CAS numbers
- produce safety data sheets (SDSs)
- carry out a risk evaluation
- develop a *exposure scenario* based on foreseeable uses of each chemical – possibly working with 'downstream' users.

Where the quantity involved is greater than 10 tonnes per year, additional obligations are placed on the manufacturer/supplier to:

- carry out comprehensive assessment of the risks from identified uses (called a Chemical Safety Assessment (CSA))
- identify the risk management measures that need to be taken.

Objectives of REACH include:

- encouraging the exchange of safety information both up and down the supply line
- preventing the putting on the market of substances:
  - where identified risks have not been adequately controlled
  - whose risks outweigh their socio-economic benefits
  - where there is a safer suitable substitute
- allowing an open market across the EU for registered substances
- placing restrictions on substances with specific risks, such as asbestos, lead, mercury, etc.

Enforcement of REACH will be by Member States' existing enforcement agencies, but an EU co-ordinating body – European Chemical Agency (ECA) – is to be established to administer the implementation of REACH. ECA will process registrations, deal with technical queries, encourage co-ordination and the exchange of information.

REACH is extensive in the minutiae of bureaucratic detail – all 850 pages of it!

The above is a very brief summary of the contents of these complex regulations and can form a general guide only. Where a specific problem arises, reference should be made to the regulations themselves.

**11.4
Transport of
dangerous goods
by road and rail**

Whenever dangerous goods are transported on the highway or by rail any incident resulting in spillage of the substance could put, not only the members of the emergency service who have to deal with it, but also members of the public at risk. This risk cannot be entirely eliminated so should an incident occur, information about the substance involved must be immediately available so the emergency services can take the correct action and hence reduce to a minimum the likely effects.

The conditions to be met when carrying dangerous substance by road or rail are contained in *The Carriage of Dangerous Goods and Use of Transportable Pressure Equipment Regulations 2004*:

These Regulations revoke most of *the Carriage of Dangerous Goods by Road Regulations 1996* and the *Carriage of Dangerous Goods by Rail Regulations 1996* and completely revoke twelve other regulations relating to the carriage of dangerous goods. They reflect EU requirements for the international transport of dangerous goods and refer extensively to the European Agreements concerning the International Carriage of Dangerous Goods by Road (ADR) and by Rail (RID) to provide some of the detailed requirements to be met.

These Regulations are in six parts and are highly complex and very detailed and only a very brief summary of the contents is given.

Part I, Introduction, includes an extensive section on interpretation and describes a range of applications within the UK and internationally. It also outlines the types of pressure vessels, equipment and tanks to which it applies but exempts certain transport arrangements of the armed forces. It allows the carrying of mixed loads provided suitable measures have been taken to prevent them coming into contact. The Secretary of State is nominated as the competent authority for matters concerning the carriage by road and rail in the UK.

Part II covers the nub of the safe operating procedures and practices and requires:

r.9 that all involved in the carriage of dangerous goods are suitably trained

r.10 the safety obligations outlined in Chapter 1.4 of ADR and Chapter 1.4 of DIR to be met

r.11 precautionary measures to be taken when consigning Class 7 (low-level radioactive) goods

r.12 the appointment of a safety adviser

r.13 the reporting of accidents and incidents

r.14 marshalling yards to have emergency plans

r.15 that only goods that have been classified should be consigned in Chapter 2.2 of ADR and Subsection 1.1.4.4 of RID

r.16 the prohibition of goods specified by RID

r.17 where there is a clash between a standard and a special condition for transporting dangerous goods, the special condition shall prevail

r.18 goods to be properly packed

r.19 special requirements to be met for the use in transport of tanks, battery vehicles, battery-wagons and multi-element gas cylinders (MEGC) whether certified or not

r.20 packages consigned for transport to be properly marked and labelled

r.21 packages for consigned goods must be manufactured, reconditioned, tested and marked in accordance with ADR and for transport by road or rail

r.22 any tank, battery vehicle, battery-wagon or MEGC must be manufactured, approved, inspected, tested and marked to comply with ADR and RID

r.23 the filling and carrying of large containers and the carrying in bulk of dangerous goods must comply with Chapters 7.1 and 7.2 of ADR and RID

r.24 the driver of any vehicle carrying dangerous goods to be trained and licensed and produce evidence if required by the police

r.25 vehicles for carrying dangerous goods on the road to comply with Part 9 of ADR.

Part III Competent authority functions

r.26 authorizes the UK competent authority (HSE) to carry out specific obligations including:
r.27 granting approval to operators
r.28 approve packaging designs
r.29 appoint persons to carry out validations
r.30 recognize and approve test methods, standards and procedures
r.31 impose conditions for the carriage of dangerous goods
r.32 issue/approve vocational training certificates
r.33 issue licenses to driver who pass a competency test
r.34 notify other State competent authorities of conditions imposed
r.35 do all other things to meet the requirements of ADR and RID
r.36 grant exemptions.

Part IV deals with the conformity of transportable pressure equipment, its marking, use, periodic inspections and maintenance. Part V refers to the carriage of Class 1 goods (explosives and pyrotechniques and their devices) and requires:

r.46 drivers transporting Class 1 goods by road to be accompanied by an attendant
r.47 Class 1 goods to be delivered 'within a reasonable length of time'
r.48 vehicles carrying Class 1 goods to ensure the security of the load
r.49 rail wagons carrying Class 1 goods to be secure and delivery made in reasonable time
r.50 all those involved in the carriage of security goods by road and rail to take steps to prevent unauthorized access
r.51 only restricted types and quantities of Class 1 goods may be transported in passenger-carrying vehicles
r.52 Class 1 goods carried in a passenger vehicle to be segregated from the passengers
r.53 care to be exercised when marshalling trains carrying Class 1 goods
r.54 copies of documents accompanying Class 1 goods to be kept by the consignor for 3 months
r.55 the consignor to be responsible for ensuring that the appropriate placards, marks and plate markings are on all vehicles carrying dangerous goods.

Part VI deals with miscellaneous matters including:

r.56 authorizing the HSE to charge fees for carrying out assessments and issuing certificates
r.57 allows a transitional defence in respect of the use of intermediate bulk containers (IBCs) and drums of more than 50 litres that they were less than 15 years old and made before 1 July 1995

r.58 allows a defence of 'due to the act or default of another person' and 'took all reasonable precautions and exercised all due diligence to avoid the offence'

r.63 refers to the revocations listed in schedule 14.

11.4.1
Danger signs

Whenever dangerous goods are being transported the vehicles are required to display danger signs to indicate the hazards from the goods being carried. These give the emergency services vital information in the event of a spillage or accident. Typical danger signs are:

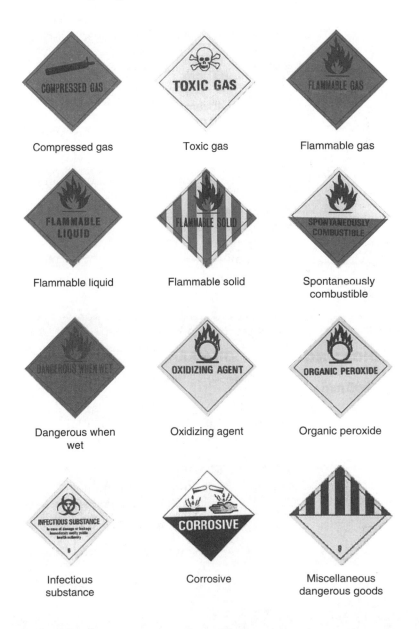

Compressed gas

Toxic gas

Flammable gas

Flammable liquid

Flammable solid

Spontaneously combustible

Dangerous when wet

Oxidizing agent

Organic peroxide

Infectious substance

Corrosive

Miscellaneous dangerous goods

11.5
Approved lists

There are two Approved Lists:

For supply – *Approved Supply List* – 'Information approved for the classification and labelling of substances and preparations dangerous for supply', Guidance on the Regulations. HSE publication L142.

For transport – *Approved Carriage List* – 'Information approved for the carriage of dangerous goods by road and rail other than explosives and radioactive materials'. HSE publication L90, also a CD-ROM (ISBN 07176 1685 1).

11.5.1
Approved Supply List

It takes account of the amendments required by CHIP 2002 and is in six parts:

Part I – for each dangerous substance gives alternative names, CAS number, index number, classification and relevant risk and safety numbers.

Part II – gives the EINECS description, index number and CAS number.

Part III – lists mixtures and mixture/specific isomers and gives index number.

Part IV – gives information on the classification of pesticides.

Part V – summarizes key information for users of the list.

The information on a substance label should include:

- hazard symbol
- risk phrase number
- safety phrase number
- EC number
- full name, address and telephone number of the supplier.

11.5.2
Approved Carriage List

This list contains information approved for the carriage of dangerous goods by road and rail other than explosives and radioactive material. It is in two parts, the first listing the substances in alphabetical order and the second listing them in UN number order. In both cases the classification information given to a substance is the same:

Column	Classification information
1	Proper shipping name
2	United Nations number
3	Classification code
4	Subsidiary hazard number
5	Emergency action code for substances carried in tanks.

This is in three parts:

(i) the number indicates suitable fire fighting means:
   1 – water jets
   2 – water fog or fine spray

3 – foam

4 – dry agent, i.e. water must not be allowed to come into contact with the substance

(ii) the first letter indicates precautions to be taken

(iii) letter E indicates that local people should be evacuated

6	Hazard identification number
7	Indication of permission to carry in tanks
8	Indication of permission to carry in bulk
9	Packaging group number indicating degree of danger:
	I – great danger
	II – medium danger
	III – minor danger
10	Indication that approval given subject to special conditions.

This classification information should be included on the hazard warning board carried on the vehicle and in the TREMCARD notes carried by the driver.

Risk phrases	**Indication of particular risks**
R1	Explosive when dry
R2	Risk of explosion by shock, friction, fire or other sources of ignition
R3	Extreme risk of explosion by shock, friction, fire or other sources of ignition
R4	Forms very sensitive explosive metallic compounds
R5	Heating may cause an explosion
R6	Explosive with or without contact with air
R7	May cause fire
R8	Contact with combustible material may cause fire
R9	Explosive when mixed with combustible material
R10	Flammable
R11	Highly flammable
R12	Extremely flammable
R14	Reacts violently with water
R15	Contact with water liberates extremely flammable gases
R16	Explosive when mixed with oxidizing substances
R17	Spontaneously flammable in air
R18	In use may form flammable/explosive vapour-air mixture
R19	May form explosive peroxides
R20	Harmful by inhalation
R21	Harmful in contact with skin
R22	Harmful if swallowed
R23	Toxic by inhalation
R24	Toxic in contact with skin
R25	Toxic if swallowed
R26	Very toxic by inhalation
R27	Very toxic in contact with skin

R28 Very toxic if swallowed
R29 Contact with water liberates toxic gas
R30 Can become highly flammable in use
R31 Contact with acids liberates toxic gas
R32 Contact with acids liberates very toxic gas
R33 Danger of cumulative effects
R34 Causes burns
R35 Causes severe burns
R36 Irritating to the eyes
R37 Irritating to the respiratory system
R38 Irritating to the skin
R39 Danger of very serious irreversible effects
R40 Possible risk of irreversible effects
R41 Risk of serious damage to eyes
R42 May cause sensitization by inhalation
R43 May cause sensitization by skin contact
R44 Risk of explosion if heated under confinement
R45 May cause cancer
R46 May cause heritable genetic damage
R48 Danger of serious damage to health by prolonged exposure
R49 May cause cancer by inhalation
R50 Very toxic to aquatic organisms
R51 Toxic to aquatic organisms
R52 Harmful to aquatic organisms
R53 May cause long term adverse effects in the aquatic environment
R54 Toxic to flora
R55 Toxic to fauna
R56 Toxic to soil organisms
R57 Toxic to bees
R58 May cause long term adverse effects in the environment
R59 Dangerous for the ozone layer
R60 May impair fertility
R61 May cause harm to the unborn child
R62 Possible risk of impaired fertility
R63 Possible risk or harm to the unborn child
R64 May cause harm to breastfed babies
R65 Harmful: may cause lung damage if swallowed
R66 Repeated exposure may cause skin dryness and cracking
R67 Vapours may cause drowsiness and dizziness
R68 Possible risk of irreversible effects

Combination of particular risks

R14/15   Reacts violently with water, liberating extremely flammable gases
R15/29   Contact with water liberates toxic, extremely flammable gas
R20/21   Harmful by inhalation and in contact with skin
R20/21/22 Harmful by inhalation, in contact with skin and if swallowed

R20/22	Harmful by inhalation and if swallowed
R21/22	Harmful in contact with skin and if swallowed
R23/24	Toxic by inhalation and in contact with skin
R23/24/25	Toxic by inhalation, in contact with skin, and if swallowed
R23/25	Toxic by inhalation and if swallowed
R24/25	Toxic in contact with skin and if swallowed
R26/27	Very toxic by inhalation and in contact with skin
R26/27/28	Very toxic by inhalation, in contact with skin and if swallowed
R26/28	Very toxic by inhalation and if swallowed
R27/28	Very toxic in contact with skin and if swallowed
R36/37	Irritating to eyes and respiratory system
R36/37/38	Irritating to eyes, respiratory system and skin
R36/38	Irritating to eyes and skin
R37/38	Irritating to respiratory system and skin
R39/23	Toxic: danger of very serious irreversible effects through inhalation
R39/23/24	Toxic: danger of very serious irreversible effects through inhalation and in contact with skin
R39/23/24/25	Toxic: danger of very serious irreversible effects through inhalation, in contact with skin and if swallowed
R39/23/25	Toxic: danger of very serious irreversible effects through inhalation and if swallowed
R39/24	Toxic: danger of very serious irreversible effects in contact with skin
R39/24/25	Toxic: danger of very serious irreversible effects in contact with skin and if swallowed
R39/25	Toxic: danger of very serious irreversible effects if swallowed
R39/26	Very toxic: danger of very serious irreversible effects through inhalation
R39/26/27	Very toxic: danger of very serious irreversible effects through inhalation and in contact with skin
R39/26/27/28	Very toxic: danger of very serious irreversible effects through inhalation, in contact with skin and if swallowed
R39/26/28	Very toxic: danger of very serious irreversible effects through inhalation and if swallowed
R39/27	Very toxic: danger of very serious irreversible effects in contact with skin
R39/27/28	Very toxic: danger of very serious irreversible effects in contact with skin and if swallowed
R39/28	Very toxic: danger of very serious irreversible effects if swallowed
R40/20	Harmful: possible risk of irreversible effects through inhalation
R40/20/21	Harmful: possible risk of irreversible effects through inhalation and in contact with skin
R40/20/21/22	Harmful: possible risk of irreversible effects through inhalation, in contact with skin and if swallowed

R40/20/22	Harmful: possible risk of irreversible effects through inhalation and if swallowed
R40/22	Harmful: possible risk of irreversible effects if swallowed
R40/21	Harmful: possible risk of irreversible effects in contact with skin
R40/21/22	Harmful: possible risk of irreversible effects in contact with skin and if swallowed
R42/43	May cause sensitization by inhalation and skin contact
R48/20	Harmful: danger of serious damage to health by prolonged exposure through inhalation
R48/20/21	Harmful: danger of serious damage to health by prolonged exposure through inhalation and in contact with skin
R48/20/21/22	Harmful: danger of serious damage to health by prolonged exposure through inhalation, in contact with skin and if swallowed
R48/20/22	Harmful: danger of serious damage to health by prolonged exposure through inhalation and if swallowed
R48/21	Harmful: danger of serious damage to health by prolonged exposure in contact with skin
R48/21/22	Harmful: danger of serious damage to health by prolonged exposure in contact with skin and if swallowed
R48/22	Harmful: danger of serious damage to health by prolonged exposure if swallowed
R48/23	Toxic: danger of serious damage to health by prolonged exposure through inhalation
R48/23/24	Toxic: danger of serious damage to health by prolonged exposure through inhalation and in contact with skin
R48/23/24/25	Toxic: danger of serious damage to health by prolonged exposure through inhalation, in contact with skin and if swallowed
R48/23/25	Toxic: danger of serious damage to health by prolonged exposure through inhalation and if swallowed
R48/24	Toxic: danger of serious damage to health by prolonged exposure on contact with skin
R48/24/25	Toxic: danger of serious damage to health by prolonged exposure in contact with skin and if swallowed
R48/25	Toxic: danger of serious damage to health by prolonged exposure if swallowed
R50/53	Very toxic to aquatic organisms, may cause long-term adverse effects in the aquatic environment
R51/53	Toxic to aquatic organisms, may cause long-term adverse effects in the aquatic environment
R52/53	Harmful to aquatic organisms, may cause long-term adverse effects in the aquatic environment
R68/20	Harmful: possible risk of irreversible effects through inhalation
R68/20/21	Harmful: possible risk of irreversible effects through inhalation and in contact with skin
R68/20/21/22	Harmful: possible risk of irreversible effects through inhalation, in contact with skin and if swallowed

R68/20/22 Harmful: possible irreversible effects through inhalation and if swallowed
R68/21 Harmful: possible irreversible effects in contact with skin
R68/21/22 Harmful: possible irreversible effects in contact with skin and if swallowed
R68/22 Harmful: possible irreversible effects if swallowed

Safety phrases   **Indication of safety precautions**

S1 Keep locked up
S2 Keep out of reach of children
S3 Keep in a cool place
S4 Keep away from living quarters
S5 Keep contents under ... (appropriate liquid to be specified by the manufacturer)
S6 Keep under ... (inert gas to be specified by the manufacturer)
S7 Keep container tightly closed
S8 Keep container dry
S9 Keep container in a well ventilated place
S12 Do not keep the container sealed
S13 Keep away from food, drink and animal feeding stuffs
S14 Keep away from ... (incompatible materials to be indicated by the manufacturer)
S15 Keep away from heat
S16 Keep away from sources of ignition – No smoking
S17 Keep away from combustible material
S18 Handle and open container with care
S20 When using do not eat or drink
S21 When using do not smoke
S22 Do not breathe dust
S23 Do not breathe gas/fumes/vapour/spray (appropriate wording to be specified by the manufacturer)
S24 Avoid contact with the skin
S25 Avoid contact with the eyes
S26 In case of contact with eyes, rinse immediately with plenty of water and seek medical advice
S27 Take off immediately all contaminated clothing
S28 After contact with skin, wash immediately with plenty of ... (to be specified by the manufacturer)
S29 Do not empty into drains
S30 Never add water to this product
S33 Take precautionary measures against static discharges
S35 This material and its container must be disposed of in a safe way
S36 Wear suitable protective clothing
S37 Wear suitable gloves
S38 In case of insufficient ventilation, wear suitable respiratory equipment
S39 Wear eye/face protection
S40 To clean the floor and all objects contaminated by this material use ... (to be specified by the manufacturer)

S41 In case of fire and/or explosion do not breath fumes

S42 During fumigation/spraying wear suitable respiratory equipment (appropriate wording to be specified)

S43 In case of fire, use ... (indicate in the space the precise type of fire fighting equipment. If water increases the risk add – Never use water)

S45 In case of accident or if you feel unwell, seek medical advice immediately (show label where possible)

S46 If swallowed seek medical advice immediately and show this container or label

S47 Keep at temperature not exceeding ...°C (to be specified by the manufacturer)

S48 Keep wetted with ... (appropriate material to be specified by the manufacturer)

S49 Keep only in the original container

S50 Do not mix with ... (to be specified by the manufacturer)

S51 Use only in well ventilated areas

S52 Not recommended for interior use on large surface areas

S53 Avoid exposure – obtain special instruction before use

S56 Dispose of this material and its container to hazardous or special waste collection point

S57 Use appropriate containment to avoid environmental contamination

S59 Refer to manufacturer/supplier for information on recovery/recycling

S60 This material and/or its container must be disposed of as hazardous waste

S61 Avoid release to the environment. Refer to special instructions/safety data sheet

S62 If swallowed, do not induce vomiting: seek medical advice immediately and show this container or label

S63 In case of accident by inhalation: remove casualty to fresh air and keep at rest

S64 If swallowed, rinse mouth with water (only if the person is conscious)

Combination of safety precautions

S1/2	Keep locked up and out of the reach of children
S3/7	Keep container tightly closed in a cool place
S3/7/9	Keep container tightly closed in a cool well-ventilated place
S3/9/14	Keep in a cool well ventilated place away from ... (incompatible materials to be indicated by manufactuer)
S3/9/14/49	Keep only in the original container in a cool well ventilated place away from ... (incompatible materials to be indicated by the manufacturer)
S3/9/49	Keep only in the original container in a cool well ventilated place
S3/14	Keep in a cool place away from ... (incompatible materials to be indicated by the manufacturer)
S7/8	Keep container tightly closed and dry

S7/9	Keep container tightly closed and in a well ventilated place
S7/47	Keep container tightly closed and at a temperature not exeeeding … °C (to be specified by manufacturer)
S20/21	When using do not eat, drink or smoke
S24/25	Avoid contact with skin and eyes
S27/28	After contact with skin, take off all contaminated clothing and wash immediately with plenty of …. (to be specified by the manufacturer)
S29/35	Do not empty into drains: dispose of this material and its container in a safe way
S29/56	Do not empty into drains, dispose of this material and its container to hazardous or special waste collection point
S36/37	Wear suitable protective clothing and gloves
S36/37/39	Wear suitable protective clothing, gloves and eye/face protection
S36/39	Wear suitable protective clothing and eye/face protection
S37/39	Wear suitable gloves and eye/face protection
S47/49	Keep only in the original container at temperature not exceeding … °C (to be specified by manufacturer)

11.6 Exposure limits

The limits for acceptable exposure to chemicals hazardous to health through inhalation for the UK are given in the HSE's publication EH40/(latest issue), 'Workplace exposure limits'. This publication is re-issued each year and contains the latest limit values based on the various criteria used.

Similar occupational exposure limits have been adopted by the EU – known as *Indicative Occupational Exposure Limit Values* (IOELVs). These EU limit values take precedence over national limits. The levels quoted in the current edition of EH 40 incorporate the IOELVs.

Occupational exposure limits are now measured as:

- *Workplace exposure limits* that are concentrations of hazardous substances in air that:
  - are unlikely to give rise to a health risk at the exposure levels quoted
  - present no serious long- or short-term health risk within time required to identify and correct any over exposure
  - can reasonably practicably be achieved and improved upon.

In the setting of limits, two groups of specialists are involved;

- Working Group on the Assessment of Toxic Chemicals (WATCH), who:
  - review the chemicals
  - consider toxicological, epidemiological and other data
  - base decisions on scientific judgement of the likely health effects

- decide if an WEL is warranted
- make recommendations on the level at which an WEL should be set.

- Advisory Committee on Toxic Substances (ACTS), who:
    - consider:
        * socio-economic implications
        * risk to health
        * cost and effort of reducing exposure
    - set an appropriate WEL level
    - make recommendations to HSC for WELs.

Exposure figures quoted are time-based:

- long-term exposure:
    - usually eight-hour period
    - averages exposure over that period (eight-hour time-weighted average [8 hr TWA])
    - over-exposure for short periods not likely to give rise to ill effects
    - any over-exposure should not exceed a duration of one hour per day

- short-term exposure:
    - usually fifteen minutes (fifteen-minute reference period)
    - quotes the level at which ill-health effects, such as eye irritation, may result from short-term exposure
    - if no short-term limit quoted, a figure of three times long-term exposure may be used.

Measures of exposure concentrations in air:

- for fumes:
    - parts per million (ppm)
    - milligrammes per cubic metre (mg/cu m)
- for dust:
    - milligrammes per cubic metre (mg/cu m).

Methods for measuring concentrations in air:

- for fumes and vapours:
    - stain detector tubes
        * single reading per tube
        * questionable accuracy
    - passive monitors (carbon discs) } need complex
    - ventilated carbon filters   chemical analysis
    - electronic gas monitors:
        * give instantaneous reading
        * continuous reading
        * incorporate warning alarm
        * can give short- and long-term integrated exposure
        * expensive

- for dusts
  - sampling pump with filter or collecting head:
    * needs accurate balance to weigh dust
    * with cellulose dust need to 'condition' filter before and after test.

EH40/(latest issue):

- lists those substances that have been given WELs
- gives methods for calculating:
  - eight-hour time-weighted average exposure for exposure to varying concentrations of the same substance
  - a method for estimating the exposure to mixtures of substances
- gives information on:
  - how the various values have been arrived at
  - an explanation of the hazards presented by particular substances such as:
    * machine-made mineral fibres (MMMF)
    * cotton dust
    * asphyxiants
    * lead
    * rubber fumes and rubber-process dust
    * flour and grain dust
    * asbestos
    * carcinogens and mutagens
    * talc
    * ferrous foundry particulate
    * wood dust
    * dust – inhalable and respirable
    * fumes

The data given in EH40 is not absolute and should be taken as a guide to the standards to be achieved. It is incumbent upon all employers to endeavour to reduce exposures to below the limits quoted.

11.7 Preventative and control measures

In the use of chemicals, there are a number of measures that can be taken to eliminate the hazard and so prevent employees being put at risk of ill-health. Where the hazard cannot be eliminated control measures must be implemented to reduce to a minimum the risk employees face from the chemicals. The primary aim of these measures should be to protect the whole of the workforce rather than the individual operator who is handling the substance.

It is helpful to follow a strategy when considering control measures:

1   **Identify** the hazards from the substances (see Section 11.2).
2   **Measure** the degree of hazard (concentration of the substance in air – see Section 11.6).
3   **Evaluate** the risk (risk assessment – see Section 3.4).
4   **Implement** preventative and control measures (outlined below).

**11.7.1
Preventative
measures**

Typical preventative measures include:

- Substitution – replacing the particular hazardous substance with one that is non- or less hazardous but which still meets the needs of the process.
- Bulk handling – by handling hazardous chemicals in bulk they can be dispensed automatically through sealed pipelines or by remote conveying. This can have the advantage of greater accuracy of dosing using electronic measuring and control techniques.
- Segregation – by separating the operator from the substances being handled. This can be achieved where manual feeding is necessary by having the substance in pre-weighed sealed bags that can be fed into hoppers that can be sealed. Alternatively, where the substance has to be weighed out, this can be done in glove boxes. Another method where the operator needs to move about, is to clothe the operator in a complete ventilated suit with an air supply from a clean source.
- Personal hygiene – much ill-health from chemicals stems from small amounts being carried on clothing or hands and being ingested when eating, drinking or smoking. There should be a ban on eating, drinking or smoking in the work area. Also employees should change their overalls and wash their hands before they eat, drink or smoke.
- Good housekeeping – accumulations of substances on the workplace floor, either as dusts or as residues left over from previous process batches, can contaminate clothing and should be cleaned up for disposal to stores or an appropriate waste bin.
- Eating arrangements – separate accommodation should be provided for the consumption of food and drink and employees should change their overalls and wash their hands before using it. Where facilities for smokers are provided they should be such that they do not interfere with others.

**11.7.2
Control measures**

Where it is not possible to take preventative measures and it is necessary to use a hazardous substance, control measures should be implemented that reduce to a minimum the degree of exposure of the operator to the substance. Typical control measures include:

- Dilution ventilation – where the concentration of a substance in air is around the maximum permitted level (WEL) it is possible to reduce that concentration to a safer level by introducing a supply of clean air. Where this is done, checks must be made of the resultant concentrations to ensure the desired reduction is achieved.
- Local exhaust ventilation – this is a system whereby the hazardous substance in the form of either dust, fume or vapour is extracted at the point of generation. This can be achieved by the use of movable extract hoods placed over the point of generation (e.g. for welding fumes, etc.) or by carrying out the process in an enclosed booth (i.e. spray booth) from which air is extracted. In both cases checks should be made to ensure there is an adequate flow of air. Also periodic examination and testing of the extraction plant is required.

- Reduced time of exposure – in certain circumstances it may be permissible to adjust the operators' work pattern so that the total time of exposure to the substance in a shift ensures that the exposure level, averaged over the shift, is well below the permitted level (WEL).
- Personal protective equipment (PPE) – this should always be considered as a last resort after the above methods have been shown to be ineffective or impracticable. It is essential that the equipment provided is suitable for the substance and the operator and does not interfere with what the operator has to do – see Section 10.2. Practical advice can be obtained from the equipment supplier and from an HSE guidance booklet 'Personal protective equipment at work. Guidance on the Regulations' Booklet no: L25.

11.8 Handling hazardous and dangerous substances

Hazardous substances can be supplied in different ways depending on the quantity involved, from the small retail package to supply in semi-bulk containers and bulk supplies by tanker. Whatever the size of the container, if a dangerous substance or substance hazardous to health is involved, the container must be labelled to identify its contents. (CHIP 2002 specifies the sizes of labels for variously sized containers.)

When handling chemicals, whether at the supply stage or during use or disposal, certain procedures should be followed to ensure operator safety. The particular procedure depends on the manner of supply or use and is considered in two parts:

1   **Bulk and semi-bulk:**

- Delivery:
  - know the safety data for the substance being handled
  - hose connections to have different threads for different substances
  - position warning sign
  - earthing link
  - wear protective clothing
  - provide eye-wash facilities
  - spillage clearance:
    * by absorbent materials
    * not flush down drain unless approved and then only when well diluted
  - follow an agreed system of work
  - over-fill:
    * have warnings to prevent
    * have procedure to draw-off surplus
  - prepare emergency procedures
  - ensure fire precautions are in place
- Use:
  - check condition of pipes for integrity
  - ensure system of work is followed
  - position semi-bulk containers in safe place

- ensure earth link is connected
- ensure materials are available to clear a spillage
- Road tankers/transport:
  - should carry hazard warning board at front and rear with:
    * HAZCHEM code
    * substance identification number
    * hazard warning sign
    * telephone number of contact for technical advice
  - driver should:
    * be properly trained
    * be provided with TREMCARD or similar:
      - for each substance carried
      - giving details of substance
    * ensure TREMCARD is:
      - kept in the cab
      - readily available
      - matches the substance being carried.

2 **Small containers and sacks:**

- Supply:
  - should be palletized
  - check for damage to packages
  - packages should be individually labelled
  - labels to:
    * identify substance
    * indicate hazards
    * give supplier's name and address
- Road transport:
  - to carry orange warning plates at front and rear
- Handling:
  - mechanical if possible
  - manual with barrows or other handling aids
  - manual as a last resort
  - procedure for clearing up spillages
- Use:
  - decant into safe, non-spill type, containers
  - earthing links for both liquids and powders
  - draw quantities for day's use only
  - measures to prevent dust from powders/granules
  - with manual loading:
    * precautions to prevent overcharging causing exothermal or uncontrolled reaction
  - check for explosion risk
  - ensure fire precautions are in place
- Storage:
  - entry to store to have HAZCHEM warning signs
  - ensure adjacent substances are compatible

- separate incompatible substances, i.e fertilizer and carbonaceous materials; acids and alkalis; etc.
- ensure every container is properly and clearly labelled
- record storage positions of each substance
- provide appropriate fire extinguishers.

Vehicle hazard warning board

Hazard warning signs

Compressed gas

Toxic gas

Flammable gas

Dangerous when wet

Oxidizing agent

Organic peroxide

Toxic

Corrosive

HAZCHEM sign for mixed storage

**11.9
Asbestos**

One of the substances that has caused much long-term ill health and many deaths is asbestos. Inhaled as a dust it can cause, among other conditions, asbestosis, which progressively reduces lung function, and mesothelioma, a malignant tumour that cannot be treated. While there is an overall ban in the EU on the use of most types of asbestos, the material has been used extensively in the past and is present in many buildings and process plants. *The Control of Asbestos at Work Regulations 2002* (CAW) are concerned with protecting employees who work in areas where asbestos may be present.

In the Regulations:

r.2 Defines:

- asbestos to cover all types
- control limit as the concentration of asbestos in air of 0.1 fibres per cc. averaged over a continuous 4-hour period.
- control measure as any measure taken to prevent or reduce exposure to asbestos.

r.3 Application extends to the self-employed as well as employers.

- Disapplies:
  - low exposure of short duration
  - removal of asbestos that is firmly linked in a matrix
  - encapsulated or sealed asbestos in good condition
  - carrying out air monitoring or collecting samples.

r.4 Defines 'duty holder' as:

- a person with maintenance responsibilities in non-domestic premises
- a person in control of premises – shared if joint tenants

and places obligations to:

- carry out a risk assessment, its periodic review and keep records
- if asbestos present, to determine the risk and identify precautionary measures.

r.5 No work to start until employer has checked for the presence of asbestos, what type and its condition assessed

r.6 Work not to start if asbestos present until it is identified, the risk assessed and precautionary measures implemented

r.7 Before undertaking any work involving asbestos, a written plan of action to be prepared including proposals to deal with the asbestos and equipment to be used

r.8 No work to be carried with asbestos until a license has been issued r.9 The enforcing authority must be notified 14 days before the start of work involving asbestos

r.10 Any employees liable to be exposed to asbestos to be given information on:

- the material and its properties
- the effect on health especially for smokers
- control levels
- safe working practices, control measures and protective equipment
- emergency procedures
- decontamination procedures
- medical surveillance.

r.11 Employers to prevent exposure of employees so far as is reasonably practicable and take measures to reduce exposure to as low a level as possible

r.12 Where control measures are provided they must be used

r.13 Control measures must be maintained in efficient working order. System of work to be reviewed periodically Records to be kept of tests of exhaust ventilation and respiratory protective equipment

r.14 Suitable protective clothing to be provided and kept clean

r.15 Suitable safety drills and emergency procedures to be in place in case of an accident

r.16 The spread of asbestos is to be prevented as far as is reasonably practicable

r.17 Where asbestos is being removed, the plant and equipment must be kept clean

r.18 Areas where work on asbestos is carried out must be designated as either *asbestos areas* or *respirator zones* and suitably identified. Entry into these areas must be controlled

r.19 Where employees exposed, asbestos concentrations in air must be monitored and records kept for up to 40 years

r.20 Air testing, both during work and for clearance on completion of work, to be to BS EN ISO 17025

r.21 Checks for the presence of asbestos to be to BS EN ISO 17025

r.22 Records of the health and medical surveillance of employees exposed to asbestos to be kept

r.23 Adequate washing and changing facilities are to be provided

r.24 Raw asbestos material and waste to be kept in sealed receptacles suitably labelled

r.25 Lists prohibited asbestos materials and operations

r.26 Prohibits exposure during spraying, removal and manufacture

r.27 Prohibits the importation in the UK

r.28 Prohibits the supply except for disposal

r.29 Prohibits the use of asbestos

r.30 Any products containing asbestos must be suitably labelled r.31 Where an exemption allows the use of asbestos, the quantity in work must be kept as low as possible

r.32 Allows the HSE to issue exemption certificates

r.33 Allows exemption from the prohibitions of rr. 25–31 in the interests of national security

r.34 Allows extension outside the UK in line with HSWA

r.35 Existing (before 2006) licences remain in effect

r.36 Refers to revocations and amendments to related legislation

r.37 Allows a defence in a prosecution of *took all reasonable precautions* and *exercised all due diligence* to avoid the commission of an offence.

These Regulations are supported by HSE publications:

L127   'The management of asbestos in non-domestic premises'

L143   Approved Code of Practice: work with Materials containing asbestos

HSG 210   Asbestos essentials task manual

HSG 247   Asbestos: the Licensed Contractors Guide

HSG 248   Asbestos: the analyst's guide for sampling. Analysis and clearance procedures.

11.10 Lead

Because of its property to resist corrosion, lead has many uses in industry and domestic life. However, it also has an ill-health effect on individuals so statutory controls on its use at work have been implemented through the *Control of Lead at Work Regulations 2002* in which:

r.2   Defines:

- action level as a blood-lead concentration of:
  - 25 µg/dl for women of reproductive capacity
  - 40 µg/dl for a young person
  - 50 µg/dl for other employees
- young person as someone who is not yet 18 years old and is not a woman of reproductive capacity
- lead to include lead alkyls, lead alloys, any compounds of lead and lead as a constituent of any substance or material which is liable to be inhaled, ingested or otherwise absorbed except vehicle exhaust fumes
- occupational exposure limits, based on 8 hour time-weighted average, for:
  - lead other than lead alkyls 15 mg/m3
  - lead alkyls 10 mg/m3

- suspension level (i.e. the level at which the person should be suspended from working with lead) as:
  - a blood-lead level of:
    * 30 µg/dl for women of reproductive capacity
    * 50 µg/dl for young persons
    * 60 µg/dl for any other employee
  - or a urinary lead concentration of:
    * 25 µg Pb/g creatine for women of reproductive capacity
    * 110 µg Pb/g creatine for all other employees
  - woman of reproductive capacity as certified by a doctor.

r.3 extends the duties under these regulations to include employees and others who may be affected by the works except master and crews of ships other than the Royal Navy.

r.4 Bans:

- the use of glazes other than leadless glaze or low solubility glaze in the manufacture of pottery
- employing young persons or women of reproductive capacity on:
  - lead smelting or refining processes
  - lead-acid battery manufacturing processes.

Employers must:

r.5
- before starting work involving lead, carry out a risk assessment and implement the precautions the assessment shows to be necessary
- review the assessment regularly and if working conditions change
- record the results of the assessment if employing 5 or more employees

r.6
- ensure employee's exposure to lead is prevented or controlled
- control can be by:
  - design of plant
  - ventilation
  - restricting exposure
  - use of PPE
- any protective equipment or clothing must be maintained in good order

r.7
- prohibit eating and smoking in lead working or contaminated areas

r.8
- maintain any control measures in good working order and review regularly the relevant safe systems of work
- keep records of examinations and tests of control measures
- provide suitable separate storage for PPE and personal clothing

r.9 • monitor the concentration of lead in air of the workplace at least once every 3 months and retain records for at least 5 years

r.10 • provide medical surveillance where the exposure is significant or where:
  – the blood-lead levels exceed:
    * 20 µg/dl for women of reproductive capacity
    * 35 µg/dl for other employees or
  – the urinary lead concentration exceeds:
    * 20 µg Pb/g creatine for women of reproductive capacity
    * 40 µg Pb/g creatine for other employees
 • keep health surveillance records for 40 years

r.11 • provide employees with information, instruction and training on:
  – the type of lead and the risk it presents
  – the relevant exposure limits
  – findings of the risk assessment
  – precautions to be taken
 • clearly identify containers or pipes containing lead compounds

r.12 • prepare suitable emergency plans and practice safety drills
 • prepare plans for the mitigation of the effects of an emergency
 • provide appropriate first aid facilities.

Hearing is something we take for granted and we subject it to all sorts of excesses. But our ears are very delicate organs and while they will accommodate a certain amount of abuse, they do eventually react and stop working properly. The following sections consider noise legislation, how the ear works and some of the measures that should be taken at work to protect the ability to hear.

When considering noise certain points need to be understood:

- noise is *unwanted* sound
- some sound is essential for:
  - communicating
  - warning
  - balance and orientation
- sounds are pressure pulses in the air
- the threshold of hearing is the lowest level of noise detectable by the ear.

Good hearing is an important feature of a good quality of life – it needs protecting.

12.1 Legislation concerning noise

There are two major pieces of legislation concerning noise, one aimed at protecting employees while they are at work and the other concerned with protecting the quality of life of the community.

12.1.1 Noise at work

The Control of Noise at Work Regulations 2005 outline actions that should be taken to reduce the ill effects of high noise levels at work and puts obligations on employers to take the appropriate action. It incorporates the requirements of the EU Noise Directive.

r.2 Defines:

- *Daily personal noise exposure,* written $L_{EP,d}$, as:
  - integrated exposure to all noise over the working day
  - can be calculated using the formula in Schedule 1
  - can be measured accurately using an integrating dose meter (dosimeter)
- *weekly personal noise exposure* as the total weekly exposure to all noise

- *peak sound pressure* as the maximum sound pressure to which an employee can be exposed
- *exposure limit value* as the daily or weekly level of personal noise exposure that must not be exceeded
- *lower exposure action value* as 80 dBA $L_{EP,d}$ or 135 dBC
- *upper exposure action value* as 85 dBA $L_{EP,d}$ or 137 dBC
- *noise* as any audible sound
- *the musical entertainment sectors* as any workplace where music, either live or recorded, is played.

r.3 places duties on employers to protect employees, and others who may be exposed, from the effects of noise

r.4 quotes exposure values:

- lower exposure action value:
  - daily or weekly personal exposure of 80 dBA
  - a peak sound pressure of 135 dBC
- upper exposure action value:
  - daily or weekly personal noise exposure of 85 dBA
  - a peak sound pressure of 137 dBC
- exposure limit values:
  - daily or weekly personal noise exposure of 87 dBA
  - a peak sound pressure of 140 dBC
- with varying exposure levels, the weekly average can be taken provided, where appropriate, the employee is given personal hearing protection.

Requires employers to

r.5 carry out a noise risk assessment to assess the level of employees' exposure, review assessment regularly and consult with employees in noisy areas

r.6 eliminate sources of noise or reduce to as low a level as possible. If noise at upper exposure action value, introduce systems of work to reduce exposure including:

- changing working patterns
- changing work equipment
- re-designing plant and plant layout
- training employees in quieter use of equipment
- ensuring good maintenance
- limiting time of exposure
- allowing suitable rest periods
- reducing exposure to below exposure limit value.

r.7 where there is exposure above the upper exposure action value, hearing protection must be provided. Such areas must be designated *Hearing Protection Zones* with restricted access and employees consulted

r.8 ensure that any measures or equipment provided to protect hearing are maintained in good working order. Employees are to use such equipment properly

r.9 where exposure and risks to hearing warrant it, medical surveillance to be provided

r.10 provide exposed employees, and affected non-employees, with information, instruction and training in the risks, precautionary measures and how to use and maintain them

r.11 allows the HSE to grant exemptions

r.12 authorizes the HSE to grant exemptions to the emergency services

r.13 authorizes exemptions in respect of the armed forces

r.14 extends the requirements to outside Great Britain

r.15 revokes earlier outdated legislation.

**12.1.2
Community noise**

Noises within a workplace can often interfere with nearby residents' enjoyment of the peace and quiet of their homes. Such noises are referred to as **community noise** and are subject to the *Environmental Protection Act 1990* which:

s.79(1)(g) Includes as a *statutory nuisance* noise emitted from premises that is prejudicial to health or a nuisance.

s.80(1) Gives powers to local authorities:

- to serve *an abatement notice* which can require the nuisance to be:
  - abated
  - prohibited
  - restricted in its occurrence or recurrence
- to put time limits for compliance.

s.80(2) Puts the onus for compliance on:

- the creator of the nuisance:
- the owner or occupier of premises emitting the noise (the employer).

s.80(3)   Allows an appeal against an abatement notice within twenty- one days of its being served.

s.80(6)   Makes failure to comply an offence subject to a fine up to £20 000.

The measurement of community noise and interpretation of the readings is very complex and is best left to specialists.

For guidance on noise at work, consult the following HSE publications:

- L108 Controlling noise at work. *The Control of Noise at Work Regulations 2005*
- HSG 109 Control of noise in quarries
- HSG 138 Sound solutions. Techniques to reduce noise at work
- INDG 362 Noise at work. Guidance for employers on *The Control of Noise at Work Regulations 2005*.

12.2
The ear

The ear is a very delicate organ capable of detecting an enormous range of sounds. Unfortunately it is taken for granted and subjected to a great deal of misuse, much of which could be avoided. By understanding how the ear works a better appreciation of the reasons for and techniques of hearing protection can be generated.

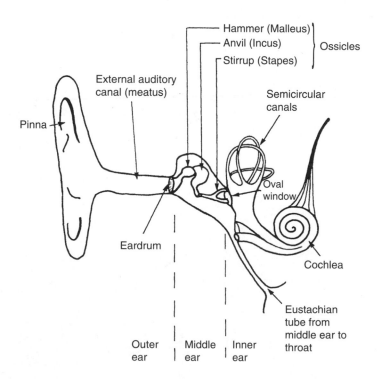

Figure 12.1 Diagram of human ear

Hearing works in the following way:

- Sound pressure pulses enter the outer ear and cause the ear drum to vibrate.
- These vibrations are transferred through the middle ear by a set of three small bones, known as the ossicles, to the oval window of the inner ear.
- The oval window transmits the vibrations to the fluid in the inner ear.
- The fluid carries the vibrations to the sensitive hair cells of the cochlea.
- The hair cells of the cochlea translate the vibrations into electrical signals which are relayed to the brain which then interprets them as sound.

**12.2.1
Effects of exposure
to excessive noise**

Exposure to excessive noise can have the following effects on:

- the ear:
  - permanent damage to the hair cells in the cochlea resulting in:
    * reduced ability to hear (*noise induced hearing loss*)
    * tinnitus (ringing in the ears)
    * a shift in the threshold of hearing with increased difficulty in hearing, particularly noticeable in a crowded room

- behaviour:
  - loss of concentration
  - loss of balance and disorientation (due to its effects on the fluid in the semicircular canals)
  - fatigue.

While the above relate to occupational noise the same effects can be experienced from leisure noise such as discos and personal hi-fis.
Hearing loss can also result from:

- blockage in the outer ear
- catarrh blocking the eustachian tube causing excess pressure in the middle ear
- a range of medical conditions, some of which may also disturb the sense of balance.

Hearing is a very precious facility that is worth taking trouble to preserve.

**12.3
Noise
measurement**

To know the extent of a noise problem it is necessary to measure the noise levels for which a number of noise level meters of varying accuracy are available and can be used. Similarly, an audiometer can be used to measure a person's hearing ability or acuity.

The **noise meter** is for measuring *noise levels*: There are three basic types:

- general purpose meter:
  - relatively cheap
  - sufficiently accurate for identifying areas with noise problems
- grade 1 instrument:
  - gives accurate readings that can be used in noise control measures
  - may include facilities for wave band analysis and integrating exposure levels
  - fairly expensive but necessary if regular noise measurements need to be taken
- precision instrument:
  - measures a range of noise functions
  - gives very accurate readings
  - often linked with recording instruments that measure noise levels over a period of time
  - very expensive and need considerable skill to use.

The **audiometer** is for measuring *hearing acuity*:

- used to measure the threshold of hearing
- indicates hearing loss
- may be manually operated or automatic in taking readings
- records the hearing ability of each ear at a series of different frequencies
- produces an audiogram (a graph of threshold of hearing for each ear at a range of frequencies)*
- test should be taken in an acoustic booth but a quiet room can be satisfactory
- medium cost but only necessary if noise is an on-going problem, otherwise use facility at local hospital.

The **dosimeter** is for measuring daily noise exposure:

- small instrument worn by employee
- consists of small recorder and a microphone which is attached to collar near the ear
- measures and records level of noise every minute of shift
- simple instrument integrates readings to give daily noise exposure $L_{EP.d}$
- more complex instruments allow detailed analysis of recorded data
- analysis requires suitable computer software and data plotter
  - very expensive
  - specialized piece of equipment best left to the specialist
- the only really accurate method of measuring the daily personal noise exposure.

*It is prudent to include an audiogram in pre-employment medical checks.

**12.3.1
Units of
measurement**

Noise measurement can be based on either 'sound power level' or 'sound pressure level'. **Sound power level** is the total sound power emitted from a body and is used in measurement of community noise, whereas **sound pressure level** is the level of noise at the point of measurement and is the more common measurement of noise level at work.

The unit of noise measurement is the **decibel**, written dB.

- The decibel is the ratio of measured noise level to minimum detectable noise level.
- It is measured on a logarithmic scale.
- The ear does not interpret noise scientifically but varies according to the frequency.
- Instruments for measuring occupational noise have modified measuring scales ('A' weighting) to match the ear's hearing characteristics, hence the occupational noise unit is dBA.
- Other weightings exist for particular applications.

The pitch of a sound depends upon its frequency, i.e. the number of pressure pulses per second, and is recorded as hertz (Hz) or cycles per second.

Figure 12.2 Weighting curves

Measurement of frequency is usually based on an octave band and recorded as 'octave band centre frequency'. In an octave band the upper frequency is twice the lower frequency. The octave band centre frequency is 1.414 × lower frequency and typical values are (in Hz):

31.5; 63; 125; 250; 500; 1000; 2000; 4000; 8000

The existence of a noise problem can be identified using a relatively simple noise meter. However, determining the measures necessary to protect against that noise may require more sophisticated and complex equipment requiring specialist knowledge to operate it and interpret the results.

12.4 Noise control techniques

In all hearing protection activities, the first consideration must be to eliminate the source of noise and hence protect all the workforce. However, it is not always possible to eliminate noise emissions completely so action must be taken to reduce emissions as far as possible. This should be done logically following a thought-out strategy.

Two approaches to a hearing protection strategy (each with a listing in order of priority) are:

1 a principles-led approach:

- elimination
  - find alternative method

- isolation
  - remove employee to less noisy area

- insulation
  - enclose noise in sound-insulated container
  - put employee in acoustic cabin

- absorption
  - line walls and reflective surfaces with sound absorbent material
  - use free-standing absorbent panels
  - hang absorbent panels from ceiling/roof

A pragmatic approach to hearing protection

- damping
  - brace or line flat-sheet metal panels to prevent drumming
  - use vibration mounts for machinery
  - use flexible connectors in pipes and ducts
  - use plastic components in machinery

- silencing
  - use silencers on exhaust from air cylinders and vacuum pumps
  - use baffles on outlets from ventilation and extraction systems
  - direct ventilation outlets away from work areas and neighbouring houses (environmental noise)

2 a pragmatic approach:

- engineer out
  - by replacement of plant
  - redesign and modification of plant
  - by altered layout of plant so areas where employees work is at acceptable noise level

- reduce noise at source
  - use of non-metallic components, i.e. plastic gear wheels, rubber bushes in linkages, etc. wherever possible
  - brace or indent metal sheets to stop drumming effect
  - use exhaust silencers, especially on exhausts from air cylinders and vacuum pumps
  - eliminate sharp bends in air and hydraulic systems to stop turbulence noise
  - eliminate electricity frequency hum in transformers – should be installed outside work area but hum can be persistent and penetrating
  - keep working parts in good order by planned maintenance
  - use fans within the maker's recommended rating to prevent air drumming
  - etc.

- contain the noise source
  - within an acoustic enclosure
  - must completely enclose noise source
  - need suitable ventilation or plant can overheat and fail

- contain the employees
  - within acoustic cabin or sound haven
  - requires suitable heating and ventilation
  - needs window or other means to view and control process
  - ban Walkmans or other personal radios

- absorb the noise
  - by use of noise absorbent materials as:
    * linings to walls
    * free-standing panels in work area
    * suspended curtains or panels.

Specialized knowledge of the material to be used and its most effective method of application is required to ensure that noise suppression and control techniques are effective. It is advisable to call in specialists.

A last resort, after all other techniques have proved ineffective, is to provide personal hearing protection which:

- must be personal issue to individual employees
- must provide sufficient attenuation (reduction in amount of noise reaching the ear) to ensure protection of hearing
- users need to get used to different level of sound that can be heard through the hearing protector.

There are two main types:

1   ear muffs:

- completely enclose each ear
- must be good seal against the head
- head band may interfere with other protective equipment
- special design for use with hard hats
- types with radio receiver in earcup are suspect since radio noise may interfere with hearing warning sounds
- can make the ears hot and be uncomfortable

2   ear plugs:

- permanent type:
  * must be fitted to individual
  * must be kept scrupulously clean or can introduce dirt in ear canal and cause inflammation preventing further use of device
- throw-away type
  * usually kept in dispensers for employees to take as required
  * one use only
  * cheap but effective
  * some require rolling to shape between thumb and fore finger for insertion in ear – hands must be clean to do this
  * do not interfere with other protective clothing or equipment.

It may be prudent to offer employees a choice of type of personal hearing protection, subject always to the chosen type providing adequate level of protection, i.e. a suitable level of attenuation and being compatible with other protective equipment that has to be worn.

A technique known as 'noise balancing', using electronic equipment to counteract the pressure pulse peaks of noise, is at an experimental stage but it is likely to be many years before it becomes commercially viable and generally available.

**12.5
Vibrations**

Noise is the transmission of vibrations through the air. However, vibrations can also be transmitted through structures, equipment and fluids. While the major effect of air-transmitted vibrations is in the ear, low-frequency vibrations can also affect certain internal body organs. Where vibrations are transmitted through equipment they tend to affect particular organs or limbs that are in contact with the equipment – such as pneumatic drills causing HAVS, specifically vibration white finger (VWF) – but can also affect internal body organs.

The Control of Vibrations at Work Regulations 2005, which incorporate the requirements of the EU Physical Agents (Vibration) Directive (2002/44/EC), have many similarities with *The Control of Noise at Work Regulations 2005*. However, vibrations are measured not by the extent of displacement (cf amplitude for noise) but by the acceleration (*A*) in m/s^2 of movement. The daily (8 hour) exposure is designated A(8) and the weekly exposure $A(8)_{week}$.

r.2 Defines:

- *daily exposure* as the quantity of mechanical vibrations a worker is exposed to over a working day averaged out for a normalized 8-hour period
- *exposure action value* is the level of daily exposure stated in r.4 at which precautionary measures should be taken
- *exposure limit value* is the level of exposure to vibrations that must not be exceeded
- *hand-arm vibration* is the mechanical vibration transmitted to the hands and arms during a work activity
- *mechanical vibration* is the vibrating movement of equipment resulting from its operation
- *whole body vibration* is the vibration transmitted to the body from a supporting surface.

r.3 places obligations on employers to protect their employees and others from the effects of vibrations.

- They will not apply until:
  - 6 July 2010 for equipment first made available for use before 6 July 2007 and
  - 6 July 2014 for whole body vibrations in the agriculture and forestry sectors.

r.4 gives exposure limit values (in units as A(8) or as m/s^2 A(8)):

- for hand-arm vibrations
  - daily exposure limit value 5 m/s^2 A(8)
  - daily exposure action value 2.5 m/s^2 A(8)
- for whole body vibration
  - daily exposure limit value 1.15 m/s^2 A(8)
  - daily exposure action value 0.5 m/s^2 A(8).

The employer is required

r.5 to carry out vibration risk assessments, record the results and review them regularly r.6 to eliminate or control the exposure to vibrations by:

- alternate working methods
- choice of equipment
- use of auxiliary equipment that reduces the vibrations
- maintaining equipment in good working order
- design of work equipment and provision of rest facilities
- training and instruction
- limiting work periods
- including rest periods in work schedules
- provision of suitable protective clothing.

r.7 to provide health surveillance where the exposure warrants it

r.8 to inform, instruct and train employees exposed to vibrations of the risks faced and the precautions to be taken

r.9 authorizes the HSE to exempt the emergency services

r.10 authorizes the HSE to exempt air transport

r.11 allows exemption of the armed forces

r.12 extends application to outside Great Britain.

Note: no defence is mentioned but since these regulations are made under HSWA, reasonable practicablitiy will apply.

The measurement and interpretation of vibrations is a complex procedure that is best left to the specialists.

Information on the risks from vibrations and the precautions to be taken is contained in HSE publications:

L 140 Hand-arm vibrations. The *Control of Vibrations at Work Regulations 2005*. Guidance on the Regulations
L 141 Whole-body vibrations. The *Control of Vibrations at Work Regualtions 2005*. Guidance on the Regulations
HSG 170 Vibration solutions. Practical ways to reduce the risk of hand-arm vibration injury
INDG 175 Health risks from hand-arm vibration: advice for employers
INDG 296 Hand-arm vibration: advice to employees.

Work equipment

With the UK's integration into the European Community the influence of Community-based legislation (Directives) on UK laws has become a major factor in current health and safety legislation. This is particularly true with regard to machinery and work equipment.

In 1989 two Directives – the Machinery Directive and the Work Equipment Directive – were adopted and their content required to be incorporated into member states' domestic laws by 1 January 1993. Both these Directives relate to plant and machinery. The UK laws incorporating these Directives are respectively:

- *The Supply of Machinery (Safety) Regulations 1992* (SMSR) – which relates to the safety standards of new plant and machinery which has been purchased since 1 January 1993. One of its aims is to promote the freer movement of goods between Member States using safety standards as the criteria.
- *The Provision and Use of Work Equipment Regulations 1998* (PUWER) – which is concerned with the use of plant and equipment that had been purchased for use at work before 1 January 1993, i.e. existing plant at that date.

These two Regulations are considered in more detail in Sections 13.1 and 13.2 respectively.

13.1
New machinery

All new machinery purchased since 1 January 1993 must comply with the *Supply of Machinery (Safety) Regulations 1992* (SMSR) which incorporate the contents of the EU Directive on the approximation of the laws of the Member States relating to machinery (Machinery Directive, no: 89/392/EEC and subsequent amendments). These Directives were aimed at removing barriers to the free movement of machinery between member states and used safeguarding as their criteria.

The Directives were drawn up under the 'new approach to legislative harmonization' whereby the main body of the Directive lays down only broad objectives to be met and qualifies these with annexes specifying particular areas or parts to be given consideration which in turn rely on harmonized (EN) standards to detail the techniques to be used as evidence of conformity. SMSR has followed this pattern.

New machinery purchased by any member state must comply with the requirements of the Machinery Directive as evidenced by the CE mark on the equipment together with supporting documentation (technical file). This applies whether a machine is purchased from a UK manufacturer or imported from another EU member state. For machinery imported from a non-EU Member State, the importer is responsible for ensuring the machine has been manufactured to EU requirements and is accompanied by the appropriate documentation (technical file).

The Regulations lay down what:

- a supplier must do before putting a machine on the market
- a purchaser can expect when he/she purchases a machine.

The main requirements of these Regulations are summarized below.

r.2 Machinery is defined as:

- *an assembly of linked components, at least one of which moves ... and which are used for the processing, treating, moving or packaging of a material*
- *an assembly of machines which in order to achieve the same end are arranged and controlled so that they function as an integrated whole ...*
- *interchangeable equipment modifying the function of a machine ...*

r.5 Excluded are equipment covered by other Directives including:

- lifting equipment
- manually operated machines
- medical equipment in direct contact with the patient
- fairground equipment
- steam boilers, tanks and pressure vessels
- nuclear plant
- radioactive sources
- firearms
- storage tanks and pipelines of fuels and dangerous substances
- transport vehicles for use on public roads
- sea-going vessels
- passenger cableways
- agricultural and forestry tractors
- machines for military or police purposes.

These Regulations do not apply to:

r.6 • machinery exported to a non-EU country

r.7 • machinery purchased before 1 January 1993

r.12 Puts obligations on a supplier to:

- design/make a machine that satisfies the relevant essential health and safety requirements (ESRs)
- prepare a *technical file* on the machine

- follow the *conformity assessment procedure* shown diagrammatically in Figure 13.2
- issue:
  - either a *Declaration of Conformity* or
  - a *Declaration of Incorporation*
- affix in an indelible manner the CE mark (Figure 13.1)
- ensure the machinery is safe

Figure 13.1 CE mark

r.14 Lists high-risk machinery as including:

- circular saws for cutting wood and meat
- sawing machines with manual-feed or loading
- hand-fed surface planers
- hand-fed tenoning and vertical spindle moulding machines
- certain types of underground machinery
- tractor power take-off shafts
- lifts for servicing vehicles
- portable chain saws
- manually loaded:
  - thicknessers for wood
  - band saws for wood or meat
  - press brakes
  - injection or compression moulding machines
  - household refuse vehicles.

For these machines the maker must:

- prepare a technical file and either:
  - submit it to an approved body for their retention
  - submit the technical file to an approved body and request:
    * verification that it meets harmonized standards
    * a *certificate of adequacy*
- or:
  - submit a technical file plus a sample machine for type testing or state where a sample of the machine might be examined.

r.17 Approved bodies:

- are appointed by the Secretary of State
- are notified to the EU
- have their name published in the *Official Journal of the European Communities*.

Documentation involved:

r.20 *Certificate of Adequacy*

- issued by an approved body
  - to certify that the machine meets EN standards
  - is presumed to conform with ESRs.

r.21 *EC Type Examination Certificate*

- issued by an approved body
  - to certify that the machine conforms with the appropriate ESRs.

r.22 *Declaration of Conformity*

- by the manufacturer that the machine meets the ESRs.

r.23 *Declaration of Incorporation*

- issued by the manufacturer
- for parts which:
  - comply with the ESRs
  - are to be incorporated into other machinery
  - cannot function independently
  - are not interchangeable.

r.24 The supplier must retain copies of the above documents and the relevant technical file until ten years after the last example of the machine was supplied.

r.25 CE mark to be:

- fixed by responsible person (who can be the manufacturer – so care needs to be exercised when buying from an unknown supplier that the machinery complies – check the supporting documentation)
- distinct, visible, legible and indelible
- in the form shown in Figure 13.1.

**13.1.1
Essential safety
requirements**

These are listed in a schedule to the Regulations and cover a range of items relevant to the safety of machinery. The headings include:

1   Essential safety requirements

    1.1 General
    1.2 Controls
    1.3 Protection from mechanical hazards
    1.4 Required characteristics of guards and protective devices
    1.5 Protection against other hazards
    1.6 Maintenance
    1.7 Indicators

2   Additional requirements for certain categories of machinery

    2.1 Agri-food machinery
    2.2 Portable hand-held and/or hand-guided machinery
    2.3 Machinery for working wood and analogous materials

3   Additional requirements for mobile machinery

    3.1 General
    3.2 Work stations
    3.3 Controls
    3.4 Protection against mechanical hazards
    3.5 Protection against other hazards
    3.6 Indications

4   Additional requirements for lifting equipment

    4.1 General remarks including test coefficients (safety factors)
    4.2 Special requirements for machinery whose power source is
        other than manual effort
    4.3 Marking

5   Additional requirements for underground machinery

    5.1 Risks due to lack of stability
    5.2 Movement
    5.3 Lighting
    5.4 Control devices
    5.5 Stopping
    5.6 Fire
    5.7 Emission of dust, gases, etc.

6 Additional requirements for equipment for lifting or moving persons

6.1 General
6.2 Hazards to persons outside the car
6.3 Hazards to persons in the car.
6.4 Other hazards
6.5 Marking
6.6 Instructions for use.

Compliance with the conditions of EU directives is through the appropriate UK laws (in this case SMSR) made as part of the UK government's commitment in Europe. However, companies that export may find that customer countries demand compliance with the word of the EU directive for the particular product regardless of whether there are appropriate UK laws or not. Compliance with the directives then becomes a matter of commercial expediency rather than statutory obligation. Any goods exported to or imported from an EU member state must carry the EC mark otherwise they may be rejected.

13.2 Safe use of work equipment

The current legislation that lays down requirements for the use of any equipment at work is the *Provision and Use of Work Equipment Regulations 1998* (PUWER). These Regulations take over the greater part of the requirements contained in the now largely defunct *Factories Act 1961* (FA) but extend to include every item of plant, equipment or tool used at work. However, the standards demanded by PUWER differ very little from those demanded by FA although there are some additional requirements.

PUWER applies to the safeguarding and safe use of all equipment, machinery and plant used in any work place. However, where machinery is concerned, any that was purchased after 31 December 1992 should comply with the *Supply of Machinery (Safety) Regulations 1992* (SMSR) (see Section 13.1) which lays down the standard of safeguarding required to enable the machinery to be sold on the open market to EU Member States but is not concerned with safe operating techniques.

PUWER cover anything provided for use at work, from a scalpel to scaffolding, a ruler to a reactor. Their requirements are summarized below:

r.2   Defines certain terms including:

* *'use' in relation to work equipment means any activity involving work equipment and includes starting, stopping, setting, transporting, repairing, modifying, maintaining, servicing and cleaning.*
* *'work equipment' means any machinery, appliance, apparatus, tool or installation for use at work (whether exclusively or not).*

Figure 13.2
Conformity
assessment procedure

r.3   These Regulations apply to:

> ... such equipment provided for use or used by an employee ... at work.

They place obligations on:

- employer
- self-employed
- any person who has control to any extent of:
  - work equipment or
  - a person at work who uses or supervises or manages the use of work equipment or
  - the way in which work equipment is used at work where the obligations extend as far as the control can be exercised.

The obligations apply whether the organization is run for profit or not. They do not apply to sea-going ships but do apply to offshore gas and oil installations.

r.4   All work equipment to:

- be suitable for the uses to which it is put
- be used only for those purposes.
- not increase risks in the workplaces

r.5   • be well maintained (such as to a programme of planned maintenance)
- have a maintenance log which must be kept up to date.

r.6   • be inspected:
  - after installation but before being operated to ensure it is safe
  - after assembly at a new site or location
  - at suitable intervals commensurate with the working environment
- inspection records to be kept
- not leave or enter an undertaking unless accompanied by evidence of the last inspection. Exceptions apply in respect of:
  - power presses and their tools
  - equipment for lifting people
  - winding equipment of mines
  - scaffolding or personal suspension equipment used in construction.
  - work equipment to which regulation 12 of the *Work at Height Regulations 2005* applies.

r.7   • be used only by trained operators
- be maintained only by authorized and trained persons.

— provide guards
for machines...

r.8 Requires employees to be given:

- adequate health and safety information about the machinery
- written operating instructions about the equipment including
  - limitations of use
  - methods of use
  - foreseeable abnormal situations
  - lessons from earlier use.
- competent and experienced supervision.

r.9 All operators and their supervisors to receive adequate training:

- in the use of the equipment
- risks associated with its use
- precautions to be taken.

r.10 Employers must ensure that items of work equipment conform with the essential requirements relevant to that equipment as included in any of the regulations listed in Schedule 1.

r.11 Means must be provided to ensure that:

- access is prevented to dangerous parts of machinery
- access is prevented to rotating stock bars
- machinery is stopped before anyone can approach dangerous parts.

Lays down a hierarchy of protective techniques in priority order:

a) fit fixed guards
b) provide other guards or protective devices such as interlocked guards, trip devices, etc.
c) use jigs, push sticks or other means that keep the operator and his/her hands clear of dangerous parts
d) the provision of information, instruction, training and supervision as is necessary.

Requires that all guards and protective devices:

- are suitable for their intended purpose
- are of good construction, sound material and adequate strength
- are maintained in effective working order
- do not create other dangers such as creating traps with moving parts
- cannot easily be by-passed or defeated
- do not allow dangerous parts to be reached
- do not interfere with the operation of the equipment
- permit maintenance to be carried out (see Section 13.4).

r.12 Lists particular hazards where special precautions must be taken to protect employees from:

- articles or substances ejected or falling from machines
- rupture or fracture of parts
- overheating and catching fire
- unexpected discharge of gases, liquids, dusts, etc.
- explosion of machine or product in it
- excludes certain processes covered by other Regulations:
  – lead
  – ionizing radiations
  – asbestos
  – hazardous chemicals
  – noise
  – wearing hard hats.

r.13 Machines, components or product at extremes of temperature to have protection against injury to employees.

Controls must be such that they:

r.14 • need a deliberate action to:
  – start
  – change condition (speed, pressure, temperature, etc.)

r.15 • include suitable controls to bring equipment to a stop
  • isolate sources of power if necessary for health and safety
  • have stop controls that take priority over all other controls

r.16 • bring the equipment to an emergency stop. These controls can be red mushroom headed switches with lock-in action, bars, levers, kick plates, pressure sensitive cable, etc. and take precedence over operating controls

r.17 • are clearly visible and labelled
• are operable without putting operator in danger
• are positioned so operator can see if anyone is in a position to be injured if the machine starts. If this is not possible, the controls to incorporate an audible or flashing light warning and delayed-start facility

r.18 • ensure that safety is not prejudiced by:
– operation of the control system
– failure in the control system

r.19 • include clearly identified and accessible means for isolating the power supply
• incorporate means to re-connect the power supply without putting the operator at risk
• incorporating isolation switches with locking-off facility.

Machines must:

r.20 • be securely attached to a base/foundation to make them stable

r.21 • be provided with suitable and sufficient lighting

r.22 • be capable of being maintained safely

r.23 • be marked to show:
– safe operating limits, pressure, weight, temperature, etc.
– hazards from equipment such as radiation, corrosive substances, heat, cold

r.24 • carry appropriate hazard warning signs [see Health and Safety (Safety Signs and Signals) Regulations 1996] and/or warning devices which must be unambiguous, easily seen and clearly understood.

Mobile work equipment should:

r.25 • not be used for carrying persons unless provided with suitable seat

r.26 • be constructed to protect the operator in the event of roll over and incorporate:
– suitable protective structure if the equipment should roll over
– restraining system (seat belt) for the operator.

r.27 • for fork lift trucks, reduce the risk of overturning to a minimum

r.28 • Self propelled mobile equipment should have:
- facilities to prevent unauthorized use
- adequate braking with back-up braking if main system fails
- good visibility using mirrors if necessary
- suitable lights if used at night or in dark places
- fire extinguisher if carrying or towing flammable materials or liquids

r.29 • Remote controlled self propelled vehicles should:
- stop automatically if they depart from control range
- have trip or similar device to prevent crushing injury

r.30 • Power take-off shafts must:
- be guarded
- have provision to prevent danger in case of shaft seizure

Additionally, a power press used for the cold forming of sheet metal should:

r.31 • exempts the presses used in certain operations.

r.32 • not be put into service until:
- it has been thoroughly examined by a competent person to ensure
  * it has been correctly installed
  * it is safe to operate
- any defects have been remedied
- guarding devices have:
  * been examined to ensure their effectiveness
  * had any defects remedied
• presses and guards to be examined by competent person:
  - every 12 months for fixed guards
  - every 6 months in other cases
  - if a mal-operation or other incident occurs that could jeopardize the operator's safety
  - to ensure defects have been remedied

r.33 • After any work on the press tools, every guard and protective device must be tested
• After 4 hours of a work period:
  - every guard and protective device must be examined and tested by a competent person
  - a certificate of compliance to be issued
• certificate of compliance should contain:
  - sufficient particulars to identify every guard and protective device
  - time and date of inspection and test
  - statement that every guard and protective device was effective

r.34 • Where a defect is found:
  – the employer must be informed
  – a report prepared giving details
  – if the defect could become a danger, a copy of report to be sent to HSE

r.35 • Reports of:
  – initial examination to be retained for 2 years
  – 4 hourly examination to be kept near the machine and retained for 6 months.

The above requirements also apply to woodworking machinery.

Guidance on these Regulations is given in the following HSE publications:

L22   Safe use of work equipment. *Provision and Use of Work Equipment Regulations 1998.* Approved Code of Practice and guidance

L112   Safe use of power presses. *Provision and Use of Work Equipment Regulations 1998* as applied to power presses. Approved Code of Practice and guidance

L114   Safe use of woodworking machinery. *Provision and Use of Work Equipment Regulations 1998* as applied to woodworking machinery. Approved Code of Practice and guidance

HSG 17   Safety in the use of abrasive wheels

HSG 31   Pie and tart machines

HSG 42   Safety in the use of metal cutting guillotines and shears

HSG 43   Industrial robot safety. Your guide to the safeguarding of industrial robots

HSG 89   Safeguarding agricultural machinery. Advice for designers, manufacturers, suppliers and users.

HSG 129   Health and safety in engineering workshops

HSG 172   Health and safety in sawmilling. A run-of-the-mill business?

HSG 180   Application of electro-sensitive protective equipment using light curtains and light-beam devices to machinery

HSG 236   Power presses. Maintenance and thorough examination

HSG 238   Out of control. Why control systems go wrong?

HSG 253   The safe isolation of plant and equipment

13.3
Safety with moving machinery

Safety in the use of machinery can conveniently be considered by looking at the hazards that the particular equipment might present. A good summary of the basic techniques and practices for ensuring safety in the use of machinery is contained in BSI's publication PD 5304 and in *Safety with Machinery* by Ridley and Pearce.

The use of machinery and equipment should be considered for all phases of its life:

- construction
- transport
- installation
- commissioning
- operation including starting up and shutting down
- setting and process change
- cleaning
- adjustment
- maintenance
- de-commissioning and dismantling.

At each phase, one or more of the following hazards may be met:

Hazard	Injury
Entanglement with the machine or material being worked on	shearing or crushing
Contact with the machine or material being worked on	friction burns or abrasions
Being trapped within the machine or between the machine and fixed material or structure	crushing
Being struck by ejected parts of the machine	physical wounds, lacerations
Being struck by ejected work material	physical wounds, lacerations
Contact with sharp edges	cuts
Being drawn in between adjacent parts	crushing
Injection of compressed air or high-pressure hydraulic fluid	emphysema
Contact with electricity	shocks and burns
Contact with hazardous chemicals	ill-health
Contact with hot surfaces	burns
Loud noise	deafness
Prolonged use of vibrating equipment	vibration white finger
Inhalation of mist and fumes	lung disorders, systemic poisoning
Inhalation of dusts	fibrosis, cancer of the lung
Ionizing radiations	skin warts, leukaemia, loss of fertility

Where any of these hazards are identified, action should be taken to eliminate or reduce to a minimum the risk faced by employees. If the hazard cannot be eliminated a risk assessment should be made to determine what safety measures should be taken. Those measures could include the provision of:

- guards
  - criteria:
    * to prevent operators reaching a dangerous part
    * with mesh guards the mesh size will determine what parts of the body can pass through and hence the distance it should be from dangerous parts
  - fixed guards:
    * requiring a tool or key to release it
    * if removable may not be replaced
    * hinged guard bolted or locked shut is more likely to be replaced
  - interlocked guard
    * can be electrical, mechanical, pneumatic or hydraulic
    * must be positive operation, i.e. switch/valve is in relaxed condition when guard is shut (negative operation – actuating spindle of switch/valve is depressed when guard is shut)
    * key-exchange system
    * captive key
    * interlock switches should be of 'fail-safe' design
  - adjustable guard
    * with parts that can be adjusted to follow the profile of the work in hand
    * most commonly used on horizontal milling machines
  - sleeving
    * the fitting of a loose sleeve round a rotating shaft
  - tunnel guard
    * must be long enough to ensure the dangerous part cannot be reached
  - automatic guard
    * moves into position automatically when the operation is initiated (i.e. press is struck on)
    * may operate by pushing the operator away from the danger area, sometimes referred to as 'push away guards'
  - control guard
    * closure of the guard actuates the machine
- fencing
  - a barrier all round a machine
  - at a distance from danger points greater than operator reach
  - has suitably interlocked access gates
  - for high-risk machines interlock should be key-exchange system
- trips
  - devices which when actuated trip the machine and cause it to stop or revert to a safe condition:
    * push bars
    * pressure mats, edgings or cable

        * photo-electric systems or electro-sensitive protective devices [floor in area covered by beams needs to be marked to prevent inadvertent actuation]
        * pull wires, normally used along the length of a conveyor
        * emergency stop switches
- systems of work
    * not normally recognized as primary means of guarding
    * usually as back-up to guarding systems listed above
    * can only be used if none of the above methods are feasible
    * should be in writing to prevent ambiguity
    * for high-risk work a permit-to-work system should be used.

Guard material can be:

- sheet metal
  - used for fixed guards or where oil splashes/spray needs to be contained
- clear plastic
  - where it is necessary to see behind the guard
    * polycarbonate is tough but soft and easily scratched
    * perspex is harder but more brittle and cracks easily
- mesh
  - allows ventilation cf. of vee-belt drives
    * Expamet needs support frame
    * weld mesh has:
      - greater strength
      - allows better visibility to see beyond guard
- grills
  - can provide considerable physical strength
  - gaps need to be narrow
  - may be visually obstructive.

There are certain necessary operations, such as setting and adjusting, that may need to be carried out with the guards open. This is permitted provided the machine can be moved, either:

- on 'limited inch' only, i.e. actuation of the operating control moves the product 3 ins (75 mm) only then stops and requires a pause before the next actuation

or:

- by 'hold-on' control where:
  - movement occurs only when control actuated
  - release of control stops the machine
  - any movement is at a pre-set crawl speed
  - use of hold-on control should isolate normal controls
  - only one 'hold-on' control is available per machine:
    * either at a number of stations round the machine, only one of which can be selected to be operational at any one time
    * by having single control on a wander lead.

On some special purpose machines such as guillotines with photo-electric safety devices or loose knife cutting machines, additional protection is provided by 'two-hand control':

- the control buttons must be at least 12 ins (300 mm) apart
- both control buttons must be actuated together
- both buttons must be released before the next actuation
- there should be a pause between sequential actuations.
- release of either button causes the machine to retract.

The above describes the main types of guarding that are in general use. However, there are many special purpose guards to suit particular applications which are perfectly acceptable provided they give the necessary degree of protection. Which type of guard is used and how it is applied must depend on the particular machine and the agreed method of operating it.

There is a great deal of guidance available through:

- British Standards
- HSE publications
- industry-based standards
- guard manufacturers.

**13.4
Safety during
maintenance**

Maintenance is one of the more hazardous types of work and particular attention needs to be paid to assessing the risks and preparing for the safe carrying out of the work. The main maintenance areas where high risks arise are:

- work on buildings and roofs at high levels
- power-driven machinery
- in confined spaces such as tanks, vats and underground chambers
- chemical plants.

Common precautions that should be taken whenever maintenance work is contemplated include:

1 Prepare specification for work to be done.
2 Assess potential risks.
3 Prepare work method statement.
4 Prepare detailed plan of work.
5 Develop safe systems of work where necessary including permit-to-work systems.
6 Check the equipment/site to ensure it is safe to start work.
7 Instruct operators and provide training where required.
8 Provide suitable work and safety equipment.
9 Prepare for unforeseen circumstances.
10 Make emergency plans.
11 Monitor that the agreed work methods are being followed.

In addition to the normal hazards of maintenance work, special hazards can arise in particular areas of maintenance work:

Buildings

- working at heights:
  - provide:
    * ladders
    * scaffolding
    * work platforms
    * safety harnesses
- fragile roofs:
  - provide:
    * roof ladders
    * crawler boards
    * edge protection – barrier rails
    * safety harnesses
  - post warning notices
- temporary power supplies:
  - ensure
    * 110-volt supply with centre tapped to earth, or
    * 240-volt protected supply, i.e. with residual current device (RCD)
- access and egress:
  - provide:
    * ladders of adequate length and in good condition
    * walkway boards over scaffolding, floor beams, etc.
    * barriers around all floor openings and floor edges
- hoists and lifts:
  - ensure they have been examined, tested and are in good condition especially in hostile working conditions
  - ensure they are adequate for the loads to be handled
  - where using hired-in lifting equipment, check that:
    * examination and test certificates are current and valid
    * the driver is fully trained and certificated
- inclement weather:
  - provide suitable bad weather clothing
  - warn operators to take extra care
  - provide facilities for drying clothing.

Machinery

- unexpected start-up:
  - isolate power supply by:
    * locking off isolator switch
    * allowing only the person who fits a locking-off padlock to remove it
    * having arrangements to cover shift change-over (a formal hand-over procedure)
  - withdrawing fuses:
    * by electrician who should certify that the machine is safe to work on
    * check by operating the machine starter controls

      * fuses should be kept in secure (locked) cupboard
      * only be replaced by an electrician
- work inside machines:
  - need to certify that the machine is safe to work on
  - where risks are high, implement a permit-to-work procedure
  - arrangements for hand-over of responsibility at shift change
  - if electrical work involved use permit-to-work system
- handling heavy components:
  - estimate weights and ensure adequate lifting tackle is available
  - ensure hoist or cranes have adequate capacity and can give a straight lift
  - if special lifting attachments required, ensure that:
    - * they are available
    - * the maintenance staff have been trained to use them
  - train staff in any special lifting techniques
  - if hired equipment is used, check:
    - * test and examination certificates for validity
    - * that the driver is fully trained and certificated
- guards:
  - do not remove until machine certified as safe to work on
  - before allowing machine back into service ensure all guards have been replaced and properly secured.
  - check operation of interlock guards
- explosive/flammable atmospheres:
  - monitor atmosphere
  - use non-ferrous tools
  - if power tools needed use those that are pneumatically-operated
  - use 'inherently safe' or external lighting
  - have fireman standing by with suitable extinguishers
  - prohibit use of naked lights
  - ban smoking
  - prepare an emergency plan
  - consider creating inert atmosphere and providing breathing apparatus
- oxygen-deficient atmosphere:
  - provide:
    - * breathing apparatus
    - * life-line and safety harness
    - * banksman at end of life-line in safe atmosphere.

Confined spaces

The Confined Spaces Regulations 1997 lays down the conditions to be met which include:

- test atmosphere in the confined space using continuous sampling electronic meter with alarm.
- continue testing atmosphere throughout whole time anyone is inside the confined space
  - if no measurable hazard:
    - * provide local ventilation from a clean source

   – if atmosphere oxygen-deficient:
   * provide breathing apparatus
   * ensure operators are fully trained in its use
   * ensure safety harness is worn and is attached to life-line
   * station banksman outside confined space to hold life-line
   * ensure the operator can be pulled from confined space in emergency, either by banksmen or using special hoist
   * prepare emergency plan
   – if atmosphere contains toxic fumes or dust:
   * follow procedure as for oxygen-deficient atmosphere (above)
- provide supplies for power and lighting at a safe voltage, 110-volt centre tapped to earth, or 24-volt
- where confined space has inlet pipes for liquids or gases:
   – ensure all inlet valves are shut and locked off [for increased safety on each feed-line, have two valves with interspace vented to drain or atmosphere].

See HSE booklet L101, Safe work in confined spaces. *Confined Spaces Regulations 1997*. Approved Code of Practice, Regulations and guidance.

Chemical plants

Major problems concern leakages from vessels and pipes:

- corrosive liquids:
   – provide 'green' protective suits
   – provide facilities for deluge washing of contaminated suits before operators change clothing
- escaping gases:
   – test atmosphere before starting work

- toxic fumes:
  - provide breathing apparatus
  - since fumes often heavier than air they will flow into depressions and pits, so atmospheres need to be tested before entering the area
- flammable or explosive gases and fumes:
  - monitor atmosphere
  - no naked lights or smoking
  - use non-ferrous tools.

While maintenance work has a bad safety record, the risks to operators can be minimized by planning each operation and ensuring the correct work methods are followed and the necessary safety equipment is provided and used.

13.5
Pressure systems

The legislation concerning pressure vessels and systems is in 2 parts, *The Pressure Equipment Regulations 1999 (PER)* dealing with the design and manufacture of pressure equipment, and *The Pressure Systems Safety Regulations 2000 (PSSR)* dealing with the use of pressure systems.

PER incorporate the requirements of an EU directive (no: 97/23/EC) and compliance with it allows free access to the EU and EEA markets. The main body of the Regulations lays down general requirements to be met and is supported by a series of schedules detailing specific aspects of the design and construction of the equipment.

1.2 Defines, *inter alia,*:

- fluid as gases, liquids and vapours in pure phase as well as mixtures of them, a fluid may contain solids
- fluids are split into 2 groups:
  - Group 1 fluids that are hazardous substances
  - Group 2 all other fluids
- pressure accessories as devices with an operational function and having pressure bearing housings
- pressure equipment includes vessels, piping, safety accessories and pressure accessories also elements attached to pressure parts such as flanges, nozzles, couplings, supports, lifting lugs and similar
- responsible person as
  - the manufacturer or his authorized agent
  - if neither is established in the EU, the person who first puts the equipment onto the market or into service
- safety accessories include:
  - pressure limitation devices such as safety valves, etc.
  - limiting devices that either maintain conditions within limits or shut the system down.

r.3 Defines pressure equipment and assemblies as those containing a pressure greater than 0.5 bar.

r.4 Refers to Schedule 1 that lists excluded equipment including:

- onshore/offshore pipelines
- water supply systems
- nuclear equipment
- oil well control equipment
- pressurized parts of machinery where the critical design feature is other than pressure, i.e. ICEs, steam engines, gas and steam turbines, compressors, pumps, etc.
- blast furnaces and associated plant
- enclosures for high voltage electrical equipment
- items with flexible casings, i.e. tyres, inflatable equipment, balls used for play, etc.
- exhaust and inlet silencers
- bottles and cans for carbonated drinks for final consumption
- vessels for transporting drinks having a pressure × volume factor of less than 500 bar.litres and a maximum pressure of less than 7 bar
- radiators and pipes of a warm water central heating system.

r.5 Excludes equipment put on the market before 29.11.99.

r.6 Allowed until 30.5.2002 for equipment designed to pre 28.11.99 UK requirements to be put on the market.

r.7 Defines pressure equipment:

- dividing vessels by the type of fluid contained, volume, maximum safe working pressure but excluding fixed or otherwise heated vessels with a volume of less than 2 litres and all pressure cookers
- grades piping accessories according to fluid contained, size of pipe and the contained pressure

Requires that pressure equipment:

- satisfies the relevant ESRs (listed in Schedule 2)
- has followed the appropriate conformity assessment procedure
- has the CE mark affixed
- is supported by a Declaration of Conformity
- is in fact safe
- any technical document shall be retained for the period specified in the document.

r.8 Requires any pressure equipment, higher temperature hot water plant to comply with the appropriate ESRs before it is put onto the market and to have followed the same procedure listed above for pressure equipment.

r.9   Refers to pressure equipment and assemblies that are:

- designed and manufactured to sound engineering practices
- accompanied by adequate instructions for use
- carry the manufacturer's identification marks
- safe

but do not carry the CE mark.

r.10 Bans the supply of pressure equipment and assemblies that are not safe.

r.11 Explains that pressure equipment and assemblies are not 'put on the market' where they are:

- for export outside the EU
- imported from outside the EU for re-export
- on exhibition at trade fairs where:
  - a notice to the effect is displayed
  - precautions are taken to protect the public.

r.12 Gives details of the classification of the equipment.

r.13 Lays down the conformity assessment procedure for pressure equipment.

r.14 Lays down the conformity assessment procedure for pressure assemblies.

r.15 Excludes experimental equipment from the conformity assessment requirement.

r.16 Deals with affixing the CE mark.

rr.17–20 Deal with notified bodies and recognized third party organizations.

r.21 Allows the charging of fees by certifying bodies.

r.22 Concerns the appointment of 'user inspectors'.

r.23 States that equipment with the CE mark and accompanied by a Declaration of Conformity shall be deemed to conform with these regulations.

r.24 Deals with enforcement action where it is suspected that the affixing of the CE mark has not complied with the laid down procedure.

r.25 Deals with the offences of contravention and failure to comply.

r.26 Lays down scales of penalties.

r.27 In the event of legal proceedings allows a defence of *took all reasonable steps and exercised all due diligence*.

Provision is made for cases where the default is due to:

- the act or default of another
- reliance on information given by another.

Schedule 2 lists the ESRs to be considerd in the design and manufacture of pressure equipment and assemblies.

The legislative requirements for the safe use of pressure vessels and plant are contained in the *Pressure Systems Safety Regulations 2000* with a supporting Approved Code of Practice 'Safety of pressure systems. Pressure Systems Safety Regulations 2000. Approved Code of Practice'. HSC publication no: L122.

These Regulations are pragmatic and recognize that differing operating conditions have differing examination requirements.

The term *pressure system* refers to the installations found in many workplaces for supplying compressed air, steam for heating and processes, various liquids, oxygen and gases for welding or burning. It includes all pipework, vessels, heat exchangers, calorifiers, boilers, etc.

The contents of the Regulations are summarized below.

r.2 Pressure systems are those that contain a **relevant fluid**, which is defined as:

- steam
- fluids (single or mixed) at a gauge pressure > 0.5 bar (7.5 psig) such as:
  - a gas
  - a liquid with a vapour pressure > 0.5 bar at its stored temperature
  - a gas that is stored by being dissolved in a porous substance and is released without heating, such as acetylene.

r.3 Refers to Schedule 1 which excludes pressure systems:

- in ships
- in weapon systems
- of a wheeled, tracked or rail-mounted vehicle
- when carried in a vehicle if they comply with international

Conventions and Rules

- any pipeline where the gauge pressure < 2 bar
- pressurized workings used in research experiments
- gas containers in motor vehicles or trailers if for propulsion, heating or cooking
- water cooling systems of engines and compressors
- portable pressurized fire extinguishers
- hand-held pneumatic tools, etc.
- diving equipment.

r.4 Pressure systems must be:

- properly designed and constructed from suitable materials
- capable of being examined safely, including internally
- provided with suitable protective devices such as pressure gauges, pressure relief valves, water level gauges, low water alarms, drains, blow-down arrangements, etc.

r.5 Vessels in a pressure system must have a fixed plate with:

- maker's name
- serial number
- date of manufacture
- standard to which built
- maximum design pressure or minimum if a vacuum vessel
- design operating temperature
- plus be marked (by painting) with:
  - the maximum working pressure
  - plant number (this helps identification during examinations and testing).

r.6 The installer of a system is responsible for ensuring it is safe to use.

A pressure system must not be used until:

r.7
- it has been examined
- its safe operating limits are known

r.8
- a written scheme of examination has been prepared by a competent person who can be an employee or a consultant. [It is usual for the company that insures the plant to assist in drawing up a scheme – they will probably be carrying out the examination as part of the service under the insurance policy.]

The scheme of examination should:

- list all protective devices
- consider all parts of the system where failure could cause danger

- state the type of examination, whether visual, internal or test
- outline preparations necessary for the examination
- lay down the frequency of examinations
- be reviewed periodically and amended as appropriate.

r.9 Examinations:

- must be carried out at the times detailed in the scheme
- recorded in a written report:
  - made by the examiner
  - handed to the user or owner
  - which states:
    * parts examined, their condition and results of the examination
    * any repairs or modifications needed and the time limit for completion
    * if condition of system warrants it, a revised safe operating pressure
    * date by which the next examination must be carried out
    * either that the system is safe to continue in present operation or give the reasons if changes are necessary.
- not use equipment until any necessary repairs have been completed

r.10 If system found to be dangerous, a copy of the report to be given to the user and sent to the enforcing authority within fourteen days. The system must not be operated until repairs have been effected.

The user of a pressure system must:

r.11 • give adequate instruction to operators to ensure:
  - they know:
    * how to operate the system safely
    * the action to take in an emergency
- check that the system is being used in accordance with instructions

r.12 • ensure it is maintained in good repair

r.13 • where a pressure system is repaired or modified, the safety of the system must not be impaired.

r.14 • keep copies of:
  - reports of examinations
  - earlier reports that refer to:
    * safe operating parameters
    * any required repairs or modifications
  - information from the supplier regarding examinations

– reports:
    * on the premises
    * readily available for inspection. They can be on disk but must be:
        – capable of providing a hard copy when required
        – secure from loss or unauthorized interference
        – be authenticated only by the examiner
        – be transferred to any new owner.

r.15 • The outlet from vessels designed to be vented to atmosphere must be kept open.

r.16 • In the event of a breach, the user can claim as a defence:
    – the offence was the act or default of another person
    – he had exercised all due deligence to avoid the offence.

13.6 Lifting equipment

The legislation concerning lifts and lifting equipment occurs in two parts – manufacture and use. This section deals only with manufacture. The use of lifts and lifting equipment is covered in section 16.

The manufacture of all lifting equipment except lifts should comply with SMSR (see 13.1), which contain the main requirements for the construction of machinery including lifts and which, for the special conditions of lifts, are supplemented by the *Lifts Regulations 1997* whose contents are summarized below:

r.2 • Defines a lift as: an appliance serving specific levels, having a car moving –
    (a) along guides which are rigid; or
    (b) along a fixed course even where it does not move along guides which are rigid (for example, a scissor lift) and inclined at an angle of more than 15 degrees to the horizontal and intended for the transport of:
        – persons
        – persons and goods
        – goods alone if the car is accessible and has controls within it.

r.3 The Regulations apply to:

• lifts permanently serving buildings and constructions
• safety components for lifts.

r.4 They do not apply to:

• cableways
• military or police lifts

- mine winding gear
- theatre elevators
- lifts in means of transport
- lifts that are part of machinery
- rack and pinion trains
- construction site hoists

r.5 • lifts purchased before 1 July 1997

r.6 • lifts purchased before 30 June 1999 that comply with the FA

r.7 • lifts to which other EU Directives apply.

r.8,9 All lifts and safety components must:

- satisfy the *essential safety requirements* (ESRs) through conformity with EN standards
- follow conformity assessment procedure
- carry the CE mark
- be accompanied by a declaration of conformity
- be safe.

r.10 Suppliers who are not manufacturers of lifts or safety components must ensure lifts they supply are safe.

r.11 • Person in charge of a building and the lift installer to keep each other informed to ensure safe and proper use of the lift.
- No unnecessary pipework or wiring to be installed in lift shaft.
- Designer to provide manufacturer and installer with all information necessary to ensure safe operation.
- The declaration of conformity must be kept for ten years.

r.12 These conditions need not be complied with if the lift is:

- destined for market outside EU
- at an exhibition when:
  - a notice stating it does not comply must be displayed
  - visitors are adequately protected.

r.13 Conformity assessment procedure for lift to include:

- final inspection
- manufacture to appropriate quality assurance scheme
- unit verification.

Conformity assessment procedure for a safety component to include:

- EU type examination
- manufacture to full quality assurance scheme.

r.14 Supplier who puts lifts and safety components on the market must ensure they conform.

r.15 Conformity assessment procedures must be carried by a *notified body*.

r.16 *Notified bodies* are appointed by the Secretary of State.

r.17 Notified bodies can charge fees for carrying out conformity assessment procedures.

r.18 A lift or safety component bearing the CE mark presumes conformity.

**13.6.1
Essential health and
safety requirements
(ESR)**

Points to be considered include:

- General:

  1 ESRs of Machinery Directive apply
  2 safety of car
  3 means of suspension and support
  4 control of loading
  5 machinery
  6 controls.

- Hazards to persons outside the car

  1 access to lift shaft only if lift isolated
  2 suitable protection at ends of lift travel
  3 gate interlocks must be effective.

- Hazards to persons in the car:

  1 cars doors must remain shut except at landings
  2 car must be fitted with fall arrester
  3 buffers provided at bottom of shaftway
  4 car must not move if arrester device not operational.

- Other hazards:

  1 motorized car and landing doors must not crush when closing
  2 landing doors must be fire-resistant
  3 counter-weights to be clear of car or other obstruction

4 cars to have:
- means of escape for trapped persons
- two-way communication system
- arrangements that allow current operation to be completed if temperature gets too high
- adequate ventilation
- emergency lighting
- communication and lighting to work in event of power failure
5 lifts capable of use in emergency must have priority controls for emergency service use.

• Marking:

1 each car must be marked with maximum load
2 car to contain instructions for emergency escape.

• Instructions for use:

1 of safety components to be in language of installer
2 for operation in language of user.

The construction industry has an unenviable record of accidents including many fatalities. While the type of work carried out does present a high level of hazard, there seems to have been an attitude endemic in the industry that it is all part and parcel of the job. Some recent major construction projects have disproved that contention and have been successfully completed without fatality or serious injury. Much of this success is due to changing attitudes particularly at senior management levels. Whether this is a reaction to new laws or whether the new laws have recognized the importance of the management aspects of construction projects, there is an increasing emphasis in legislation on the wider responsibilities of senior managers and of other off-site people such as architects who play a crucial role in determining on- site safety. These new legislative requirements are embodied in the *Construction (Design and Management) Regulations 2007* (CDM).

These Regulations, which revoke the earlier CDM 1994, now embrace the contents of the *Construction (Health, Safety and Welfare) Regulations 1996* which is also revoked. A feature of the 2007 Regulations is that the role of the planning supervisor has been changed to that of co-ordinator and re-titled CDM co-ordinator.

A major development aimed at improving safety on construction sites has been the coming into effect of the *Work at Height Regulations 2005* that deal with the hazards associated with such work and which cover specifically both falls of persons and falling objects.

The contents of both these Regulations are outlined in the following sections which also include advice on a number of common practices within the construction industry.

**14.1
CDM**

CDM are the initial letters of the *Construction (Design and Management) Regulations 2007*, which place responsibilities for safety on construction sites on everyone involved, from the designer and the client to principal and sub-contractors.

The main requirements of these Regulations are outlined below.

Part 1   Introduction

r.2   Defines:

- client as anyone seeking someone to or carrying out himself a project

- construction site includes any place where construction work is being carried out
- construction phase plan as a document covering health and safety arrangements, site and any special rules
- construction work includes the preparation for, carrying out and maintenance work involved in a project
- contractor as anyone who carries out construction work
- designer as anyone involved in design work for a project
- health and safety file containing records and information likely to be needed in future work on the project and includes a Health and Safety File prepared under CDM 1994
- pre-construction information covers all the information designers and contractors may need to complete the project safely
- principal contractor as the contractor with overall control of the construction work
- project includes planning, designing, managing and other work until the end of construction
- structure as any erection whether above ground or below and includes formwork, falsework and scaffolding
- notifiable as any construction work lasting more than 30 days or involving more than 500 person days.

r.3 these Regulations apply in Great Britain and outside as far as HSWA applies

Part 2   outlines management duties and obligations

r.4 no one to offer or accept an appointment unless satisfied as to their competence

r.5 all concerned in a project to co-operate in the safe carrying out of the work

r.6 all concerned in a project to co-ordinate their activities

r.7 all concerned in a project to follow the general principles of prevention in carrying out the work

r.8 there should be only one client for each project

r.9 client responsible for making arrangements that ensure safe working

r.10 at pre-construction stage, client must provide information to designers and contractors to enable them to ensure that work can be carried out safely by all involved or affected by it r.11 designers must ensure that their designs eliminate hazards and reduce risks to ensure safety during:

- construction work
- cleaning of windows, translucent walls and roofs
- maintenance of permanent fixtures
- use of structure as a workplace.

r.12 for designs from outside Great Britain, the person who commissions them or the client must ensure they comply

r.13 contractors must:

- ensure client knows his duties
- plan his work so it can be carried out safely
- allow sub-contractors time to plan their work
- give those under his control training:
  - site induction
  - risks identified by risk assessment
  - precautions to be taken
  - site rules
  - emergency procedures
  - identity of person in charge
- ensure security of the site
- provide suitable welfare facilities.

Part 3 places duties for notifiable projects

r.14 Client to appoint in writing:

- a CDM co-ordinator
- principal contractor.

In the absence of such appointments, client assumes responsibilities.

r.15 Client to provide CDM co-ordinator with pre-construction information as r.10

r.16 Client to ensure work does not start until contractor has produced a *construction phase plan* and provided welfare facilities

r.17 Client to provide all information for CDM co-ordinator to create a *health and safety file* and keep file available at end of project

r.18 Designer not to start detail design work until CDM co-ordinator appointed then provide him with adequate information about the design r.19 No contractor to start work until:

- he is provided with the names of the CDM co-ordinator and the principal contractor
- he has access to the construction phase plan
- notice of project has been given to HSE or rail regulator.

Every contractor to provide principal contractor with:

- information on risks from his work
- names of sub-contractors he appoints
- information on any accidents or incidents.

Contractors to:

- comply with directions of principal contractor
- obey site rules
- follow construction phase plan
- notify principal contractor of need to change construction phase plan.

r.20 CDM co-ordinator to:

- give advice to client on action to comply with CDM Regulations
- ensure health and safety arrangements are made and implemented ensuring co-operation between all those involved in project and that they apply the general principles of prevention
- liase with principal contractor on:
  - contents of health and safety file
  - information needed for the construction phase plan
  - any design developments likely to affect planning and management of construction work
- identify and collect pre-construction data
- provide such information to designers and contractors
- ensure designers meet their obligations
- assist in co-operation between designers and contractors where changes are made
- prepare a *health and safety file* and pass it to client at end of project.

r.21 • notify HSE, or rail regulator as appropriate, of project
- notification to be signed or approved by client.

The health and safety file could include:

- a brief description of the project
- pre-construction information prepared by the client
- copy of the *construction phase plan*
- information provided by the designer

- list of contractors and sub-contractors
- information from main and sub-contractors about, or copies of their:
  - safety policies
  - PL (Public Liability, or third-party) and other relevant insurance cover
  - method statement of how the work will be carried out
  - test and examination certificates of any lifting equipment, whether hired in or owned
  - certificates of competence (licences) of crane drivers, truck drivers, scaffolders, etc.
  - notes of meetings between client and contractor.

r.22 Principal contractor to:

- plan, manage and monitor the project without risk to health and safety
- liase with CDM co-ordinator
- ensure welfare facilities are adequate
- draw up site safety rules
- give contractors time to plan and prepare their work
- consult with contractors concerning their work
- ensure contractors provide information to CDM co-ordinator for the health and safety file
- display a copy of the notification to HSE or rail regulator
- ensure the security of the site
- ensure every site worker gets:
  - site induction training
  - health and safety training from his employer
  - any special health and safety training required by his job.

r.23
- Prepare a construction phase plan at start of project and review and update it regularly
- ensure the implementation of the construction phase plan.

r.24
- Develop co-operation with workers on health and safety matters
- consult workers on matters affecting their health, safety and welfare
- make information affecting their health and safety available to workers.

Part 4   lists health and safety duties on construction sites

r.25 Contractors and anyone who control any work on the project to comply with rr.26–44 and anyone working on the site must report any defect or danger r.26 There must be safe access to and egress from places of work which must be kept safe

r.27 Site to be kept in good order, clean and secure

r.28 New and existing structures to be safe

r.29 A written plan of demolition to be prepared to prevent danger

r.30 Explosives to be stored, handled and used safely

r.31 Precautions at excavations to be taken to prevent:

- collapse of excavations and persons being trapped or buried
- people, equipment or materials from falling into excavations
- adjacent ground being overloaded
- work in excavations until it and equipment inspected
- work in excavations until any faults remedied.

r.32 Cofferdams and caissons to be:

- properly designed and constructed
- equipped with means of escape
- inspected before use, at start of each shift and after any incident
- not used until any identified fault has been corrected.

r.33 Reports of inspections to be:

- completed and delivered within 24 hours
- delivered by end of shift if a fault found
- kept available for inspection
- retained until 3 months after the end of the project.

r.34 • Energy distribution systems, air, electricity and hydraulic, to be located to prevent danger, checked and clearly identified
- Electric power cables to be run clear of areas of risk or isolated. Overhead cables to have suspended warnings
- Checks to be carried out to locate underground services.

r.35 Where there is a risk of falling into water steps to be taken to:

- prevent such falls
- minimize the risk of drowning, i.e life buoys and safety boats
- provide rescue equipment
- prevent overload of water transport.

r.36 Traffic routes to be:

- separated from pedestrian routes
- adequate for the size and number of vehicles
- provided with barriers at pedestrian outlets
- provided with separate pedestrian exits at loading bays
- indicated by suitable signs
- checked regularly
- well maintained.

r.37 Vehicles on site to:

- be controlled to prevent unintentional movement
- be fitted with device to give warning of movement
- be loaded and driven in a safe manner
- not carry passenger unless suitable seat fitted
- allow driver to leave vehicle while it is loaded/unloaded
- not be driven to risk falling into excavations, pits, waterways, embankments, etc.

r.38 Measures to be taken during construction to prevent injury from:

- fire or explosion
- flooding
- asphyxiating atmospheres.

r.39 Procedures to be in place for dealing with emergencies with account taken of:

- type of work being carried out, size and layout of the site
- equipment being used
- number working on site
- substances being used
- need to train all on site
- need to carry out regular practice drills.

r.40 Emergency exit routes to be provided that:

- lead to a safe area
- are kept clear
- if necessary have emergency lighting
- are suitably identified.

r.41 Fire precautions to include:

- fire-fighting equipment
- fire detection and alarm systems
- regular inspections and tests
- manual fire fighting equipment to be easily reachable

- training in fire fighting techniques
- in high-risk areas, special training (and fireman in attendance)
- fire fighting equipment and alarm points to be clearly identified.

r.42 Every place of work and access way to have wholesome air supply. If ventilating plant necessary it must have failure alarms

r.43 The temperature of indoor work areas must be reasonable. For work out-of-doors suitable foul-weather clothing to be provided

r.44 Places of work and access routes to be adequately lit. The colour of artificial lighting not to interfere with safety signs. If necessary for safety, emergency lighting should be provided

r.45 Non-employees do not have the right of civil action for a breach of these regulations except as regards:

- welfare facilities
- unauthorized entry
- obligations placed by r.26–44.

r.46 The enforcing authority for fire prevention and fire safety matters is the HSE

r.47 Deals with transitional provisions

Schedule 1 lists the particulars to be notified to the HSE or Rail Regulator

Schedule 2 deals with welfare facilities and requires the provision of:

- adequate number of conveniences for men and women
- suitable washing facilities with soap, towels, etc.
- adequate supply of wholesome drinking water
- changing rooms with secure lockers
- suitably furnished rest areas with facilities, where relevant, for nursing mothers.

Schedule 3 lists detail items to be included in an inspection report

Schedule 4 lists the legislation that is revoked

Schedule 5 lists amendments to other legislation

The main thrust of these Regulations is at the management of construction projects to ensure that the work is properly planned, hazards and risks anticipated, and that adequate and suitable facilities are available to protect the health and safety of anyone working on the site.

HSE publications L54 and HSG 224, which are based on the 1994 Regulations, do give general advice on good construction practices. They have been superceded by L144 *Managing health and safety in construction. Construction (Design and Management) Regulations 2007. Approved Code of Practice and guidance.*

14.2
Work at Height

The Work at Height Regulations 2005 are aimed at reducing the enormous toll of injuries received as a result of work people falling from work equipment. They are complementary to the *CDM Regulations* and place great score on the planning of work at height, the provision of suitable equipment and competency of those involved in its use.

r.2 Defines:

- *access* and *egress* to include ascent and descent
- *fragile surface* as a surface that could fail if a reasonable load was applied to it
- *ladder* to include fixed ladders and step ladders
- *work at height* as any place above or below ground level and access to and egress from it except by a permanent staircase
- *working platform* as any platform used for work or as a means of access and includes scaffolding, cradles, trestles, gangways, gantries and stairways.

r.3 Extends its application to areas covered by HSWA but excludes normal shipboard activities and certain dock operations

r.4 Requires every employer to:

- plan properly work at height
- provide appropriate equipment
- supervise the work
- ensure it is carried out safely taking account of weather conditions.

r.5 Everyone involved in the organization, planning and supervision of work at height to be competent

r.6 Steps should be taken to avoid the risk of falling by:

- carrying out a risk assessment and acting on it
- if reasonably practicable carry out the work at ground level
- if not, take measures to prevent people falling
- if preventative measures are not possible, reduce the distance and consequence of fall and give additional training.

r.7 Work equipment must:

- give collective protection
- be suitable for the work being carried out
- allow passage without risk
- and take account of:
  - working conditions
  - where the work is to be carried out

- how far access and egress is
- likely height of a fall
- emergency evacuation
- extra risks posed by other equipment.

r.8 Places requirements for specific protective measures including:

- hand rails, etc. to be to Schedule 2
- scaffolding to Schedule 3
- air bags, netting, etc. to Schedule 4
- personal protection to Schedule 5
- ladders to Schedule 6.

r.9 Fragile surfaces should be identified by a warning notice and employees warned.

- If it can be done elsewhere, work should not be carried out on or near fragile surfaces or
- if that is not reasonably practicable suitable access platforms with hand rails to be provided or
- the height and consequences of a fall to be minimized

r.10 Materials and objects should be stored so they will not fall. Waste materials to be tipped in a safe manner.

r.11 Areas where there is a risk of falling or being struck by falling, items should be identified and barriered off.

r.12 Work at height equipment to be inspected:

- if assembly critical, before use
- if subject to adverse environment, regularly and after abnormal conditions
- work platforms over 2 m high, within 7 days of erection
- lifting equipment, on arrival and departure from site by a competent person and certificate kept for 3 months.

r.13 Fall protection measures to be checked before each use

r.14 Duties are placed on employees to:

- report any defects or dangers from equipment
- use equipment as instructed.

Special requirements relate to climbing and caving

r.15 Gives the HSE powers to issue exemption certificates

r.16 Exempts the armed forces

Schedule 1 Requires work places, access and egress to:

- be stable and strong enough
- be on firm footings

- be of sufficient size to allow work, access and storage of materials
- have integral floor with no gaps
- be used properly and kept maintained
- prevent moveable parts moving during work.

Schedule 2 Protective guard rails, barriers, etc. to be:

- large and strong enough and securely fixed
- located to prevent falls
- guard rails to have:
  - top rail at 950 mm
  - intermediate rail
  - toe board
- mounted on secure structure
- only have openings at access to ladder or stairway
- kept in position except for special tasks and replaced immediately task finished.

Schedule 3 Part 1

The structure for working platforms must be:

- stable while being erected, altered and during use
- mounted on a firm footing
- prevented from moving by attaching to secure structure
- mobile structures must have locking to prevent inadvertent movement.

Working platforms to be:

- of sufficient size to allow work, movement and storage of materials
- provided with integral floors that are non-slip
- arranged to prevent trapping against adjacent structures
- of adequate strength for the loads to be supported.

Part 2

Scaffolding to be:

- properly planned and stressed
- built to recognized configuration
- erection and dismantling to be planned
- appropriate to the work being undertaken
- not used until erection complete
- partial erection to carry warning notices
- erected, altered and dismantled under supervision of competent person.

Schedule 5 Fall protection methods

All-person devices such as nets, inflated bags, etc. to:

- be securely attached
- have adequate clearance.

Personal fall arrest devices to:

- have a force-limiting component
- have a clear fall zone
- arranged to ensure the supporting line cannot be cut.

[Note: if actuated, may need an emergency recovery procedure to be in place]

Schedule 6 Ladders should:

- only be used if it is safe to do so
- rest on a firm base
- be footed
- secured at its upper end
- protrude above the platform far enough to form a handhold
- have a platform if the rise is more than 9 m
- if extending, ensure sections lock together.

Step ladders should only be used for low risk short duration work

Schedule 7 Reports of inspections should include:

- name and address of client
- location of equipment
- description of equipment
- date and time of inspection
- safety matters
- corrective action needed
- name of inspector.

14.3 Construction safety

Whenever construction work is to be undertaken, if the work is properly planned, consideration is given to some of the hazards faced and the appropriate precautions taken, accidents and damage can be prevented. This applies whether you are a client having the work done or a contractor doing the work. However, once having agreed precautionary measures it is essential that the work is monitored to ensure the precautions are implemented. The contractor should supply *all* plant and materials needed for the contract.

Typical of the points to be considered are:

- Falls from heights:
  - working platforms:
    * must be provided with hand rail at 950 mm, intermediate rail at 500 mm and toe board 150 mm high
  - for roof work, provide:
    * edge protection
    * safety harness
    * crawler boards if work is on a fragile surface (such as asbestos sheeting)
  - ladders must:
    * be in good condition
    * be inspected regularly
    * have feet placed on solid compacted ground or suitable spreader board
    * be lashed at top end to structure. (If this is not possible post someone at bottom to 'foot' the ladder)
    * if used as access, project at least 1m above platform served unless safe handhold provided
    * be set no steeper than 75°, i.e. 1 unit out for every 4 units rise
    * if longer than 9 m be provided with an intermediate platform.
- Falling objects:
  - hard hats must be worn
  - when material stored at high level, ensure platform is:
    * wide enough to allow access past storage position
    * strong enough to support the weight
    * provided with hand rails, toe board and netting or similar to retain materials
- Trenches and excavations deeper than 1.2 m (4 ft) should:
  - have sides:
    * battered (sloping) or
    * benched (stepped) or
    * properly shored under supervision of competent person
  - be inspected very day
  - be far enough from existing buildings not to affect foundations
  - have edges protected to prevent:
    * people falling in
    * vehicles falling in
  - have ladder for access and egress
  - have arrangements for removing water
- Temporary wiring
  - should be secured to structure, not left hanging in loops
  - connections should be properly made, not taped
  - should be kept for minimum period then removed
  - should be 110-volt, centre tapped to earth (see Section 17.3)
  - if 240-volt, should be protected by residual current device (RCD)
- Power tools
  - should be preferably 110-volt
  - should be subject to regular PAT testing, including leads, sockets, etc.

- Materials
  - follow supplier's safety instructions
  - provide PPE (gloves, goggles, face masks, etc.) as appropriate
  - where solvents are used (in adhesives, paint, etc.) ensure work area is well ventilated
- Housekeeping
  - ensure site kept tidy
  - do not allow build-up of rubbish – it is a fire and health hazard
  - rubbish to be removed not burnt on site
  - dust and fumes to be kept to a minimum
- Noisy equipment
  - used only during 'social' hours (7am to 7pm)
  - provide hearing protection to operators and those working nearby
- Overhead power lines
  - indicate by lines of bunting or flags
  - post warning notice
- Underground services
  - check with local authority and gas, water, electricity and telephone companies before excavating
  - check for service runs with suitable instrument
- Scaffolding
  - must be erected by competent certificated (CITB) erectors
  - must be inspected:
    * before use
    * every week
    * after damage or alterations
- Asbestos
  - check site for its presence
  - check type – local analytical chemist can do this

- get atmosphere monitored to determine concentrations
- decide action, i.e. remove, seal or encase
- if removing, arrange for an approved contractor to strip it out
- use specialist contractor
- Pneumatic tools
  - breakers, chisels, etc.
  - risk of vibration white finger (VWF)
  - if diagnosed move worker to other work
  - check with supplier of tools for availability of attachments to reduce the vibration effect (insulated handles, etc.)
  - provide hearing protection for operator and those working nearby
- Mobile cranes
  - ensure test/examination certificates are current
  - ensure driver is properly trained and certificated (CITB)
  - use outriggers when lifting
  - ensure outrigger feet rest on suitable base plates
  - ensure base plates are on solid compacted ground
  - allow room for swing of counterbalance weight
  - use a banksman to assist driver with slinging and lifts
  - if crane hired in:
    * check test/examination certificate of crane and associated equipment
    * check driver is trained and certificated
    * if in doubt refuse to accept the hire
- Welfare facilities
  - each contractor should provide their own but can share by arrangement
  - facilities include toilets, washrooms, canteen, first aid, etc.
- Site arrangements
  - should be made known to all contractors:
    * routes to be taken by workmen and vehicles
    * security arrangements
    * fire precautions and alarm
    * site facilities available to the contractor
- Equipment
  - each contractor should provide their own
  - if contractor wants to use another's equipment:
    * agree in writing
    * borrower to give written statement that equipment is in good order
    * if equipment needs driver, either owner provides one or contractor's driver must prove competence (by training certificate)
    * borrower must sign indemnity accepting responsibility for any damage caused by or to equipment during period of hire/loan
- Pits, openings, and platform edges to be provided with a substantial barrier
- Use of local services:
  - agree with site agent before work starts:
    * which services
    * for how long
    * what charge to be made.

Establishing good communications with site agent and between contractors, through the nomination of individuals to be the points of contact, will enable many of the day-to-day problems to be sorted as the work progresses and ensure it is carried out in a much safer manner.

14.4 **Employing** **contractors**	At some stage in their existence most companies employ contractors to carry out work on their premises. This work can range from building, electrical installation, installing and commissioning a new plant, etc., through to cutting the grass and cleaning windows.

While the contractor is on the premises the occupier has both statutory and common law responsibilties for his, contractor's, health and safety and that of his employees, but may not have any control over how those employees work and behave. By the same token, the contractor has obligations to ensure that the occupier's employees are not put at risk from the way he carries out the work covered by the contract.

The safe and successful carrying through of a contract depends on good relations and communications between the occupier and the contractor. This is helped by having well defined and understood conditions in the contract.

Occupier's responsibilities include:

- ensuring contract includes conditions requiring the contractor to comply with all relevant statutory health and safety legislation
- informing the contractor of any special hazards on the site
- providing training for the contractor, his employees and sub-contractors in the precautions to be taken
- providing copies of the local safety rules and requiring the contractor, sub-contractor and their employees to comply with them
- identifying the area where the work is to be carried out and over which the contractor will have control
- appointing a responsible person to be the point of contact with the contractor
- not allowing own equipment to be used by the contractor without prior written agreement
- checking that the contractor has adequate Employers Liability (EL) and Public Liability (PL, or third-party) insurance cover. EL cover is mandatory; PL cover should match likely losses (minimum of £10 m)
- providing safe access for the contractor's vehicles and employees
- informing the contractor of site arrangements, i.e. hours of work, services (air, water, etc.), welfare facilities that he can use (toilets, canteen, first aid), fire prevention arrangements, emergency procedures, etc.
- informing contractor of procedure for obtaining clearance to excavate and, where known, the routes of underground services.

Reciprocally, the contractor's responsibilities include:

- having a safety policy and ensuring it is implemented on the site
- having adequate insurance cover
- preparing a programme for carrying out the contract

- preparing method statements for each phase of the work
- ensuring all relevant statutory legislation is complied with
- organizing the work to enable it to be carried out in a safe manner
- informing his own and sub-contractor's employees of special hazards on the site and training them in the precautions to be taken
- employing only skilled workmen and providing competent supervision
- informing the occupier of any equipment, materials or processes he may use that could give rise to a risk to the health of the occupier's employees and seek the occupier's permission before using them
- where relevant, informing the enforcing authority of the intended commencement of the work
- taking all necessary steps to ensure that the occupier's employees are not put at risk from the manner in which the work is done
- providing all the equipment, materials and services needed to carry out the work safety and efficiently
- instructing his and sub-contractor's employees in the site emergency procedures
- instructing his own and sub-contractor's employees of the site facilities they can use
- checking for underground services before excavating
- providing any necessary fire-fighting equipment for the part of the site under his control.

Within the above responsibilities, typical subject matters to be considered include:

- area of the premises over which contractor given control
- procedure for the transfer of control for that area
- means and routes of access
- use of site services; electricity, air, water, gas, etc.
- use of site welfare facilities; canteen, first aid, toilets, etc.
- site arrangements:
  - hours of work
  - access, traffic routes, speed limits
  - parking
  - emergency procedures
  - smoking areas
- site safety rules
- safe systems of work
- restrictions on use of occupier's equipment
- contractor's safety policy
- contractor's insurance cover and any restrictions or limitations
- liaison, communications and points of contact
- special training
- statutory requirements applicable to the site and the contract
- use of hazardous and dangerous materials, substances and equipment
- hand-over procedure at the end of the contract.

By working closely with contractors, the work can be expedited and the site made a safe place to work.

Sources of advice and guidance:

- HSE publications available from HSE Books
  - HSG 33 Health and safety in roof work
  - HSG 47 Avoiding danger from underground services
  - HSG 107 Maintaining portable and transportable electrical equipment
  - HSG 141 Electrical safety on construction sites
  - HSG 144 The safe use of vehicles on construction sites
  - HSG 149 Backs for the future. Safe manual handling in construction
  - HSG 150 Health and safety in construction
  - HSG 159 Managing contractors. A guide for employers
  - HSG 168 Fire safety in construction work
  - HSG 185 Health and safety in excavations. Be safe and shore
  - HSG 189/1 Controlled asbestos stripping techniques for work requiring a licence
  - HSG 213 Introduction to asbestos essentials
  - HSG 224 Managing health and safety in construction (based on 1994 Regulations and due to be revised for 2007 Regulations)
  - HSG 227 A comprehensive guide to managing asbestos in premises
  - L102 A guide to the *Construction (Head Protection) Regulations 1989*
  - Managing contractors – A guide for employers (an open learning booklet) (ISBN 0 7176 1196 5)
  - A guide to managing health and safety in construction (ISBN 0 7176 0755 0)

Many of the above are summarized in short free publications (1 copy free but a charge for additional copies)

- 'Construction Hazard and Safety Handbook' by R W King and R W Hudson, Butterworth-Heinemann

---

14.5
Access equipment

The relevant legislation is *The Work at Height Regulations 2005* which, in r.7, requires priority to be given to collective protection from falls and, in r.8, outlines the requirements for equipment used for work at heights and includes scaffolding, working platforms, access equipment, etc. However, since scaffolding is used so extensively it will be dealt with in some detail.

Scaffolds must:

- be erected by a trained and competent person who is certificated by CITB or approved training organization
- be inspected by a competent person:
  - before being taken into use
  - every seven days

Figure 14.1 Typical putlog scaffold

- after inclement weather
- after changes to the scaffolding
- report of each inspection kept available in site office
- be of good construction, suitable and sound material and adequate strength
- not contain defective materials
- be securely supported, properly braced and, if relying on building for stability, rigidly attached to the building
- have standards or uprights vertical or slightly inclined towards the building with feet on adequate base plates
- have ledgers horizontal
- have well secured putlogs
- have hand rail 950 mm high, intermediate rail at 500 mm and toe boards 150 mm high on gangways, stairs and working platforms where there is a risk of falling.
- gangways and working platforms to be at least 600 mm wide
- scaffold boards to be laid so that:
  - no gaps large enough to cause injury
  - joins level so no risk of tripping

- supported every 1 m for boards 32 mm thick
- supported every 2.6 m for boards 50 mm thick
- access ladders to project at least 1 m above platform level and be properly secured

...free standing towers must be stable...

- if partially erected or dismantled, display a prominent warning notice and have access blocked
- be properly maintained and kept in good condition at all times
- timber parts of good quality wood, stripped of bark and left unpainted (so defects are not hidden).

Free-standing access towers must:

- be no higher than three times length of shortest side
- be constructed to be stable (use of cross braces)
- be used only on firm level and even surface
- have access ladders secured inside tower
- if mobile (fitted with wheels):
  - not be moved with anyone on the tower
  - be moved by applying force at or near the base
  - have wheels locked when in use.

'Snorkel' and 'bean stalk' type passenger lifts must:

- be used only on firm level ground
- use outriggers if provided
- follow maker's recommendations for load and radius
- not be overloaded.

Sources of information and guidance:

- HSE publications:
    HSG 32 Safety in falsework for in-situ beams and slabs
    HSG 33 Health and safety in roofwork
    HSG 141 Electrical safety on construction sites
    HSG 144 Safe use of vehicles on construction sites
    HSG 150 Health and safety in construction
    GS28 Safe erection of structures; pt 3 Working places and access
    PM28 Working platforms on fork lift trucks
- BS 5973 Code of Practice for access and working scaffolds and special scaffold structures in steel

---

**14.6
Safety in
demolition**

Within the construction industry, which has a notoriously bad safety record, one of the more dangerous operations is that of demolition. However, the work can be carried out in relative safety if certain practices and rules are followed.

The legislation covering demolition is very brief and is contained in r.29 of *The Construction (Design and Management) Regulations* 2007 which says:

Demolition or dismantling
29. (1) *The demolition or dismantling of a structure, or part of a structure, shall be planned and carried out in such a manner as to prevent danger or, where it is not practicable to prevent it, to reduce danger to as low a level as is reasonably practicable.*
     (2) *The arrangements for carrying out such demolition or dismantling shall be recorded in writing before the demolition or dismantling work begins.*

It should be noted that the requirement is to prevent danger if at all *practicable*, i.e. technically possible without consideration of cost and, only if that cannot be achieved, is the lesser requirement of reducing the danger *so far as is reasonably practicable* allowed. A written plan of the work must be prepared before any demolition starts. This can conveniently be considered in three phases:

1 Planning the work
2 Health hazards
3 Working methods.

**14.6.1
Planning the work**

In the initial stages, consideration to be given to:

- the type of structure or building
- its condition
- the existence of services – whether live or dead
- presence of residual hazardous substances from previous use, such as asbestos, flammable and toxic materials
- effect of or on adjacent buildings

- access to site for plant
- storage of demolished materials
- removal of debris and spoil
- programme of work:
  - sequence of working
- methods of work:
  - prepare method statements
  - systems of work including permits-to-work
- selection of contractor:
  - competent
  - experienced in the type of demolition
  - provide competent supervision.

**14.6.2
Health hazards**

Hazards to health can arise from the presence of:

- gases, fumes, vapours and dusts which may be asphyxiant, toxic, corrosive or carcinogenic and cause ill-health through:
  - inhalation
  - ingestion
  - absorption through the skin or open wounds
- flammable liquids and materials.

Typical substances met in demolition include:

- lead
- asbestos
- cement dust
- silica
- residues from previous processes on site
  - manufacture, use, handling, storage, etc. of known or unknown materials.

Appropriate precautions must be taken and personal protective equipment provided.

**14.6.3
Demolition methods**

Methods of demolition include:

- Hand demolition:
  - follows reverse order from construction
  - dangers from falling debris so no working allowed under parts being demolished
  - poor access over rubble and to parts to be demolished
  - lack of working platforms at parts being demolished
  - if chimney used as debris chute, clear base regularly
- Mechanical demolition
  - demolition ball:
    - * vertically
    - * swinging in line with jib $\Big\}$ heavy duty crane must be used
    - * slewing jib

- pusher arm
- hydraulic hammer
- power grapple and shears
- grabs
- Where mechanical demolition used:
  - roof trusses hand-demolished to bearer plate
  - hand demolition to 1 m wide from adjacent buildings
  - stability of adjacent building to be checked and maintained
  - only machine and banksman within 6m of part of building being demolished
  - machine should have clear working space of 6m from building
  - the building should not be entered once demolition has started
- Deliberate collapse
  - structure pre-weakened then use either explosives or wire rope pulling.

Note: this technique requires knowledge of the strength of building structures and should only be carried out by competent operatives under strict supervision.

- Explosives
  - to be used only by specialists with knowledge of:
    * firing charges and sizes
    * blast protection necessary
    * area of exclusion zone
    * firing programme
    * how to handle misfires
    * precautions to prevent predetonation from radio and similar interference
    * warning signs and policing of exclusion zone
    * security arrangements for storage of explosives
- Wire-rope pulling
  - not to be used on structures over 21 m high
  - steel rope with minimum diameter of 38 mm to be used
  - winch or pulling vehicle positioned at least two times building-height from building
  - winch or pulling vehicle to have steel cab or cage to protect operator
  - exclusion zone to extend from building and on each side of the rope a distance at least three-quarters of the distance of winch or pulling vehicle from building
  - if pulling fails:
    * building or structure must not be approached or entered
    * other mechanical means to be used to complete demolition.

Fire should **never** be used as a method of demolition.

Special techniques that should be described in the work method statement are required for:

- pitched roofs
- filler joist floors
- floors supported by steel beams

- jack arches
- reinforced concrete floors
- framed structures
- in-situ cast reinforced concrete:
  - beams
  - columns
  - walls
  - suspended floors and roofs
- pre-cast concrete structures, floors and wall panels
- pre-stressed concrete

Demolition is a highly dangerous process and should only be undertaken by specialists with the appropriate qualifications and experience. Guidance can be obtained from:

- BS 6187, Code of Practice – Demolition
- HSE publication HSG 150, Health and safety in construction
- HSE Guidance Notes: GS29 Health and safety in demolition work

---

**14.7
Safety with
excavations**

Many building and construction projects require excavations, from a simple trench for pipe-laying to deep foundations for a major building. Any work at levels below the surface of the ground can give rise to risks from collapse or slippage of the ground, falls of persons or materials, and flooding. By following simple but well established techniques, those risks can be reduced to a minimum.

This section is not concerned with tunnelling which is a specialized operation that should be carried out only by organizations competent and experienced in that type of work.

Legislative requirements concerning excavations are contained in r.31 of CDM 2007 and include:

- all practicable steps (technically possible without consideration of cost) to be taken to prevent:
  - collapse of excavation
  - falls of materials
  - a person being buried or trapped
  - persons or equipment falling in
- *suitable and sufficient steps* to prevent surrounding ground being overloaded
- excavations with supports or battering to be inspected by a competent person:
  - before use
  - at start of work shift
  - after event affecting the strength of the excavation
  - after a fall of material
- no work to proceed until the competent person is satisfied of the safety of the excavation.

(a) Battering the sides

Figure 14.2 Safe
trenching methods
without the use of
timber

Maximum 1.2 m (4 ft)

(b) Benching the sides

It is accepted that excavations less than 1.2 m (4 ft) deep do not present a serious risk to those working in them. In any excavation of greater depth suitable precautions must be taken.

Typical hazards from excavations:

- collapse of sides
- materials falling onto those working in the excavation
- people or vehicles falling in
- undermining adjacent structures and buildings
- damage to underground services
- asphyxiating fumes and gases
- flooding.

14.7.1
Prevention

Considering each of the hazards listed above prevention methods include:

- Collapse
  - benching
  - battering
  - shoring.
  If shoring:
  - timber to be used must be:
    * inspected
    * in good condition
  - be erected by experienced workmen under the control of a competent person experienced in the techniques
  - be inspected before excavation is used
  - can also be:
    * trench boxes with hydraulic struts
    * metal-sheet piling

- Material falling in
  - store any spoil, plant or materials clear of the edge of excavation
  - fit toe boards along edge of opening to prevent loose material falling in
- People falling in
  - work people
    * provide substantial barrier
  - if public can approach:
    * fence off completely
- Vehicles falling in
  - provide suitable barrier or stop-log which should be painted to stand out
  - if need to tip into excavation provide stop-blocks at appropriate distance from edge
- Adjacent structures
  - check foundation or footing of structure
  - if risk of weakening them shore the walls before digging
- Services:
  - include gas, water, electricity, telephone
  - establish runs:
    * from local plans
    * by using pipe/cable locator (Note: these devices do not detect plastic pipes)
    * hand-dig trial holes to confirm location of services
    * hand-dig in vicinity of services
    * when exposed provide support for service carrier
    * report to service provider any damage to service
- Gases and fumes
  - from engines, boilers, LPG cylinders, solvents, etc.
  - check atmosphere in excavation before each shift
  - duct fumes away from excavation, or
  - provide adequate fresh air ventilation
  - fumes can lie in excavation for some time after source removed
- Flooding
  - provide suitable ladders for escape
  - provide pump of adequate capacity to keep excavation dry.

Inspections play a critical role in accident prevention. Inspections should be carried out on:

- all excavations:
  - before being used
  - at the beginning of every shift
  - after falls of materials
  - if excavation has been damaged

These inspections do not have to be recorded but it is prudent to do so. However, where the competent person is not satisfied about the safety of an excavation he must prepare a report giving the particulars listed in Schedule 3 of the Regulations. He must give a copy to the contractor who must keep the report available for inspection until 3 months after completion of the work.

The precautions outlined above can be implemented with very little disturbance to the work but could prevent the enormous amount of disruption that inevitably follows an accident.

Guidance is given in: HSE publication HSG 185 Health and Safety in excavations, and BS6031 Code of Practice for earthworks.

Manual handling

Over one third of all reported lost-time absences are attributed to injuries caused by manual handling. This problem is common across Europe and has resulted in a Directive aimed at reducing this toll. In the UK the contents of the Directive are contained in *The Manual Handling Operations Regulations 1992* (MHOR).

A number of practical techniques have been developed to facilitate the manual handling of loads and hence reduce the toll of injuries and some of these are considered in this chapter.

15.1 The Manual Handling Operations Regulations 1992

These Regulations are aimed at reducing the appalling toll of injuries from manual handling. The requirements of the Regulations are summarized below.

r.2 Defines:

- manual handling operations as *any transporting or supporting of a load (including lifting, putting down, pushing, pulling, carrying, or moving thereof) by hand or by bodily force*
- load as including any person and any animal
- excludes illhealth caused by content of loads.

Puts duties on employers regarding their employees; the self-employed are responsible for themselves.

r.3 The Regulations do not apply to ship-board activities of sea going ships.

r.4 Employers should avoid manual-handling operations that could put employees at risk of injury [i.e. use mechanical handling wherever possible].

Where manual handling is unavoidable, employers should:

- carry out a risk assessment of the handling operation covering:
  - the tasks
  - the loads

- the working environment
- capacity of the individual to do the work
- other factors such as the effect of protective clothing, etc.
- reduce the risks from manual handling to a minimum
- provide information on:
  - the weight of loads to be handled
  - off-centre and out-of-balance loads
- review the risk assessment when conditions change.
- take account of:
  - the employee's:
    * physical abilities
    * clothing, footwear, etc.
    * knowledge and training
    * whether especially at risk
  - the results of risk assessments
  - the results of any health assessments.

r.5 Employees are required to follow instructions aimed at reducing the risk from manual handling

r.6 Authorizes exemptions for the armed forces

r.7 Extends application outside UK as far as HSWA applies.

Under HSWA, employers are required to provide suitable training to employees and this extends to include training in the various manual handling techniques – see Section 15.2.

Advice and guidance on safe manual handling is given in the following HSE publications:

HSG 60 Upper limb disorders in the workplace
HSG 115 Manual handling. Solutions you can handle
HSG 119 Manual handling in drinks industry
HSG 121 A pain in your workplace? Ergonomic problems and solutions
HSG 149 Backs for the future. Safe handling in construction
HSG 171 Well handled. Offshore manual handling solutions
HSG 225 Handling home care
L23 Manual handling. *Manual Handling Operations Regulations 1992. Guidance on the Regulations*

15.2 Safe manual handling

Of the enormous toll of injuries attributed to manual handling by far the greater number (two-thirds) are strains and sprains. By knowing the mechanics of the body as they apply to lifting and handling and by following some simple rules the toll of these accidents can be much reduced.

15.2.1
Definitions

- **Manual handling** – any transporting or supporting of a load by hand or bodily force whether lifting, putting down, pushing, pulling, carrying or moving.
- **Sprain** – severe wrench or twist of a ligament or muscle of a joint causing pain and swelling of the part resulting in pain and difficulty in moving.
- **Strain** – injury done to a limb or part of the body through being forcibly stretched beyond its proper length.

15.2.2
The lifting
mechanics of
the body

Understanding the mechanics of the parts of the body that are involved when a load is lifted or carried is basic to developing techniques and practices to ensure the muscles are not overloaded.

Figure 15.1 shows a simple line diagram of the skeleton carrying a load (W) at a distance (y) from the spine.

Figure 15.2 shows a diagrammatic detail of the vertebrae of the spine with the discs, the spinal cord and the back muscles attached to a horn-like projection from each vertebrate. The backbone pivots about the discs between the vertebrae with reactionary force applied by the muscles. The whole of the load that is lifted is taken by the spine.

Figure 15.1 Body mechanics of lifting

The bending moment about the spine from the load $= W \times y$

This is resisted by the moment of the pull of the muscle times its distance from the vertebrate $= P \times r$

Therefore for balance $W \times y = P \times r$

Thus the load on the spinal muscle $P = W \times y \div r$

If r is small compared to y then the load on the spinal muscle is many times greater than the load being carried. For example if a load of 10 kg is carried at elbow's length, i.e 400 mm from the spine, and the distance of the spinal muscle from the centre of the disc is 20 mm then the load on the spinal muscle is:

$$10 \text{ kg} \times 400 \text{ mm} \div 20 \text{ mm} = 200 \text{ kg} \text{ (450 lbs or 4 cwts)}$$

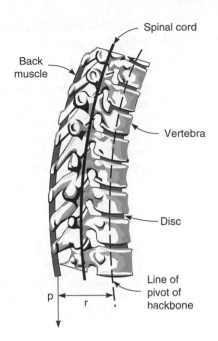

Figure 15.2 Mechanics of the backbone

Figure 15.3 Suggested maximum loads at various distances from body

Thus to prevent back strain the load carried should be:

• kept as low as possible
• carried as near to the body as possible.

Other muscles, in the arms and wrists for example, are attached to bones near to their pivot point and similarly carry many times the load of the weight being lifted.

Figure 15.3 shows recommendations for maximum loads at various distances from the body and at differing heights.

15.2.3
Preventing strain

Where a load has to be moved there are a number of simple steps that can be taken to prevent back and other muscle strain:

- wherever possible the load should be moved by mechanical means
- if not, an assessment made of the risks involved in manually handling it
- the load should be split down into manageable sizes
- each part should be within the lifting capability of the employee
- if still too heavy, assistance should be sought to help with the lift
- employees should be trained to:
  - follow the correct lifting techniques including kinetic handling by:
    * placing feet firmly
    * gripping load firmly
    * using arms and legs to lift – raising load in stages, resting it on knees or thighs while the position of the back is changed
    * not trying to lift with the back, i.e. not changing position of the back when lifting or holding the load
    * taking the strain slowly when picking up the load
    * not snatching the load
    * holding the load as close to the body as possible
    * not twisting the body when lifting or carrying a load
  - estimate the weight of loads
  - ask for assistance if the load is too heavy
  - wear appropriate protective clothing:
    * gloves to protect hands from splinters
    * safety shoes to protect feet should load be dropped
  - report any difficulties to the supervisor.

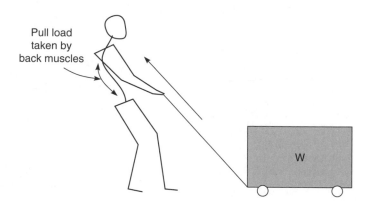

Pull load taken by back muscles

W

Figure 15.4 Dynamics of pulling a load

When moving loads on a trolley:

- Pulling
  - the whole strain is taken by the back muscles – see Figure 15.4
  - if load moves suddenly it can run on and injure employee's feet
- Pushing
  - at optimum height of push (about mid-chest height) the strain passes down the backbone – see Figure 15.5

Push load
taken by
back bone

W

Figure 15.5 Dynamics
of pushing a load

– pushing at too high a level puts excessive strain on stomach muscles
   and can cause strain of shoulder muscles and hernia
– pushing at too low a level is ineffective.

Typical problems encountered with loads include:

- weight
- shape
- size
- compactness or looseness of packing
- off-centre loads.

Where continued repetitive movement or handling causes pain – called
repetitive strain injury (RSI) or work related upper limb disorder
(WRULD) – the rhythm or routine of the work should be changed or
arrangements made for a rotation between different jobs. If condition
persists seek medical advice.

For further guidance, consult HSE publication no. L23 'Manual handling, guidance on the Regulations', which gives advice on the interpretation of the Regulations and examples of a number of techniques that can
be used.

16 Mechanical handling

Mechanical handling refers to any powered means for moving or lifting loads. The more common means are:

- lifts, including hoists
- cranes, including lifting tackle
- conveyors
- powered trucks, including driverless trucks.

All new equipment, whether operated by power or manually, must comply with:

- The *Supply of Machinery (Safety) Regulations 1992* (as amended) (SMSR)
- additionally for lifts, with *The Lifts Regulations 1997* (LR) which lay down the standards of construction to be met.

Once put into use, all equipment must comply with:

- The *Provision and Use of Work Equipment Regulations 1998* (PUWER)
- and additionally for lifting equipment, with *The Lifting Operations and Lifting Equipment Regulations 1998* (LOLER) in respect of both the condition of the equipment (and its means of protection) and the methods of operation.

The uses of different types of lifting equipment are considered in more detail in this chapter.

16.1 Lifting

Lifting equipment covers any equipment used in raising or lowering any load, including people. It ranges from the massive tower cranes used on construction sites to a dumb waiter in a restaurant, and includes a bath hoist in a nursing home, a scissor lift, elevating conveyors, lorry tail lifts, a passenger lift, fork lift trucks, etc.

The *Lifts Regulations 1997* define a lift as:

- a car or platform serving specific levels
- moving along fixed guides or
- along a fixed course without guide rails (i.e. scissor lifts)

- whose path is more than 15 degrees above the horizontal
- intended to transport:
  - people
  - people and goods
  - goods alone if the controls are accessible from outside or to a person in the car

This chapter is not concerned with the details of the design or construction of lifting equipment but only with the methods of work and procedures to be followed for ensuring the equipment is safe to use.

The relevant legislative requirements are contained in LOLER whose main contents are summarized below.

r.2 • Defines:
- lifting accessories as any equipment for attaching to the load to enable it to be lifted
- lifting equipment as any equipment used in the raising or lowering of loads
- load to include a person
- work equipment as any machine, tool or other equipment used at work

r.3 • Applies to all lifting equipment for use at work and places duties on the employer, self employed and anyone who has control of lifting operations but does not apply to the lifting equipment of a ship.

r.4 • Requires that lifting equipment must be strong enough for the loads to be handled and stable when handling them.

r.5 • Lifts for carrying people must:
- prevent anyone using it from being injured or
- if work has to be carried out from the cage, ensure the person is adequately protected;
- have anti-fall arresters or enhanced safety factors on the suspension rope with daily inspection by a competent person;
- ensure that anyone trapped can be safely released

r.6 • lifting equipment must be installed so that:
- the risk of the load or equipment striking anyone is a minimum
- the load cannot drift, fall freely or be released unintentionally
- it is safe
- persons cannot fall down shafts or hoistways

r.7 • Lifting equipment and accessories must be marked with its safe working load

r.8 • Lifting operations must be:
- properly planned

>   – adequately supervised
>   – carried out in a safe manner
>
> r.9 • Lifting equipment must be thoroughly examined:
>   – before being put to use for the first time unless accompanied by an EC declaration of conformity
>   – periodically to a programme of examinations
>   – after a potentially damaging incident
>   – after long periods out of service
>   • lifting equipment should not be moved between undertakings without evidence of previous examination
>
> r.10 • Any defects found during an examination:
>   – to be notified to the employer and
>   – owner if leased or hired
>   – followed by a written report. If defects give rise to imminent danger, a copy of report giving prescribed information to be sent to the enforcing authority
>   • Defective equipment not to be returned to use until repairs effected
>
> r.11 • Copies to be kept available of:
>   – EC declaration of conformity
>   – periodic examinations.

Advice and guidance on complying with these Regulations is given in the HSE publication L113 Safe use of lifting equipment.

Continuous operating lifts of the paternoster type are exempt from having gates but must have edge trips to stop movement in the event of anyone becoming trapped between part of the cage and the cill or lintel of a landing. Their speed of movement is restricted.

Escalators and travelators, while strictly not lifts, transfer people between fixed points. Precautions to include:

- skirting to prevent trapping between tread and balustrading
- combs to tread at entry and exit
- adequate clear area of floor at entry and exit
- clearly indicated emergency stops devices
- regular and proper maintenance including cleaning the area of the return track
- periodic thorough examination.

With scissor lifts:

- access must be prevented to the scissor mechanism by:
  - setting the base in a pit
  - having roller screens, shutters or similar between the table and base wherever mechanism is exposed
  - providing fences round the non-operating sides.

When servicing a scissor lift, the table should be retained in the elevated position by chocking the moving wheels in the base. Do *not*, repeat *not*, attempt to chock the table since this can tilt upwards and allow the scissor mechanism to close.

16.2 Cranes

Cranes are lifting equipment that have no restraint on the direction of movement. Typical cranes are:

- chain blocks — single direction of movement
- mono rails — two directions of movement
- overhead travelling cranes
  mobile cranes
  tower cranes — three directions of movement.
  jib cranes

The construction of all new cranes bought since 1 January 1993 should comply with the *Supply of Machinery (Safety) Regulations 1992* (as amended) and carry the 'CE' mark. Cranes should be used and maintained in accordance with LOLER. Both these Regulations refer to:

- *lifting machinery* rather than cranes
- *lifting accessories* – meaning components or equipment not attached to the machine and placed between the machinery and the load or on the load in order to attach it to the crane hook
- *separate lifting accessories* – meaning accessories that help to make up or use a slinging device, such as eyehooks, shackles, rings, eyebolts, etc.

To ensure safe use of cranes:

- driver, slinger and banksman must be trained and certificated
- driver should be assisted by a banksman who directs the driver in the movement of loads
- there should be only one banksman to direct crane driver
- a slinger, or if no slinger the banksman, slings or attaches the load to the crane hook
- lifting equipment and tackle must be properly stored
- there should be regular visual inspections of ropes and rope slings – this should be done every time they are used
- all standard hooks should have a safety catch – some 'C' hooks, specially designed to prevent displacement of the load without a safety catch, are still in use
- overload alarms should be fitted wherever feasible
- cut-out switch should be fitted between the sheaves of the lifting hook and crane crab to prevent jamming and over-straining of the lifting ropes (it is expensive to replace a lifting rope).

Additional precautions needed when using mobile cranes include:

- use of outriggers
- ensure outrigger feet are:
  - on solid base boards
  - on well compacted level ground
- employ only trained, certificated and authorized drivers
- allow room for swing of counterbalance weight particularly when near buildings
- if working near overhead power lines, position warning flags or bunting and post warning notices
- if crane hired-in, check:
  - examination and test certificate is up to date – if not, do not allow on site
  - driver supplied by hirer is fully trained and certificated by CITB – check licence
  - the crane is suitable for the work to be done
  - the hirer has adequate insurance cover (min £10 m), either general public liability (PL) or special cover for the job.

Construction site lifting equipment is subject to the above requirements and, because of the hostile environment in which they have to work, may require more frequent examinations and checks.

16.3 **Conveyors**	Conveyors are used to carry a flow of goods or materials, and can present a number of hazards. There are two broad types of conveyor: roller conveyors and belt conveyors. A common hazard is that of trapping, either between adjacent rollers or under the belt.

16.3.1
Roller conveyors
	• free running type:
	  – normally gravity feed
	  – no trapping hazards except at entry from belt conveyor (Note: in Figure 16.1 the first roller C is arranged to lift out of its housing)
	• power-driven rollers:
	  – if all rollers driven there are no intake traps
	  – if alternate rollers driven, trap can be formed between driven and idler rollers – provide strip guard across conveyor at these points

16.3.2
Flat-belt and trough conveyors
	• guarding to be provided at intakes to:
	  – belt drive from the motor
	  – head pulley for at least 1 m from intake
	  – tail pulley for at least 1 m from intake
	  – snub or tensioning pulleys
	  – rollers where belt changes direction
	  – rollers under feed hoppers where side of hopper prevents belt lifting off rollers
	  – return belts below conveyor to provide protection when sweeping floor (see Figure 16.2)

Figure 16.1
Preventing free
running roller trap

- guarding of long straight runs can be by trip wire which must:
  - have a lock-out switch at each end
  - be within easy reach of anyone trapped by belt
  - be arranged so that pulling in either direction trips the conveyor
  - when tripped, have controls arranged to require reset before conveyor can be restarted
  - keep switches clean and clear of debris
  - have its function checked regularly
- guarding by distance fencing must:
  - be high enough and positioned so belt cannot be reached
  - have interlocked gates to permit access for adjustment of, repairs to and cleaning of the conveyor
  - provide reasonable gap under fence to allow for sweeping the floor

With storage systems based on areas of flat or gravity roller conveyors, adequate longitudinal and cross walkways must be provided to enable operators to get to loads and push them safely (see Figure 16.3).

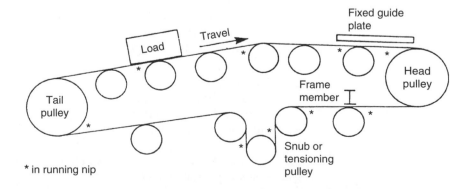

Figure 16.2
Diagrammatic layout
of belt conveyor
showing in-running
nips

Figure 16.3 Walkways between and across roller conveyors

**16.4
Powered Trucks**

This section does not cover road vehicles but only those trucks used solely within a works premises.

A common requirement for the safe use of all powered trucks is that the drivers are fully trained and competent.

**16.4.1
Training**

The training of drivers should:

- be by qualified instructors – check their credentials. If a training centre they should be members of the Industrial Truck Training Association
- be held in-house or at accredited training centre
- cover:
  - driving theory, especially for fork trucks with rear-wheel steering and other specialized trucks with multi-wheel steering
  - use of basic controls
  - practical driving techniques
  - practical experience in the workplace
  - test of theoretical knowledge and practical competence.

Drivers who successfully complete training should be given a certificate of competence (licence).

**16.4.2
Fork-lift trucks**

Candidates for fork-lift truck training should:

- be over eighteen years of age
- have a medical check before training starts
- have stereoscopic vision. If monocular (one-eyed) they should be referred for a medical check on their vision
- not be colour-blind
- if epileptic or diabetic, be referred to doctor for medical check
- not be known drug users.

Safe operating techniques for fork lift trucks include:

- trucks should be fitted with back and overhead guards
- driver should be aware of the outward swing of the back of the truck when turning due to back-wheel steering
- powered steering prevents spinning of steering wheel if road wheel hits an obstruction – if manual steering the steering wheel should be recessed for the steering knob
- carrying of passengers prohibited unless purpose-built seat fitted
- trucks must not be used to raise persons unless fitted with a special-purpose working platform, for details see HSE guidance note PM28
- floor must be in good condition and level
- when travelling up or down a slope the load must be towards the top of the slope
- must not be driven with the forks elevated, either loaded or unloaded
- ignition key or starter card must not be left in an unattended truck
- priority right of way must be given to pedestrians
- loads must be within maker's recommended limits
- lifting chains should be examined every six months
- pedestrian accessways should be separate from truck lanes
- when charging battery electric trucks:
  - hydrogen given off by battery
  - battery cover should be removed or perforated cover used to allow hydrogen to escape
  - high-level ventilation should be provided to charging area to disperse hydrogen
  - provide water supply and eye-wash bottles in case of acid splash
  - when replacing batteries ensure lifting hooks do not make contact with a battery terminal.

Additional special training should be given to drivers of rough terrain lift trucks.

Most common causes of accidents with lift trucks are:

- turning at speed
- turning across a slope
- turning with the load elevated
- skylarking by unsupervised youngsters.

Drivers guilty of any of these practices should have their authorization to drive (licence or certificate) suspended until further training completed. For guidance on the safe use of lift trucks:

- British Industrial Truck Association's booklet 'Operator's Safety Guide for Powered Industrial Trucks' and 'BITA's Operator Safety Code for Rough Terrain Lift Trucks', 10th edn.
- HSE guidance booklets.
- HSG 6 Safety in working with lift trucks
- HSG 113 Lift trucks in potentially flammable atmospheres
- HSG 136 Workplace transport safety. Guidance for employers
- HSG 221 Technical guidance on the safe use of lifting equipment offshore
- L113 Safe use of lifting equipment. *Lifting Operations and Lifting Equipment Regulations 1998*. Approved Code of Practice and guidance
- L117 Rider-operated lift trucks – operator training, Approved Code of Practice and guidance
- Guidance Note no. PM 28 Working platforms for fork lift trucks.

16.4.3 Pedestrian operated trucks

Usually the operator walks in front of the truck where the main hazard is over-run of the vehicle with consequent injury to driver's feet.
   Precautions include:

- over-run trip on control handle
- fitting of 'cow catcher' guard to front wheels
- wearing of safety footwear.

16.4.4 Automatic guided vehicles(AGVs)

Used where frequent transfers made between discrete fixed points. Normally follow a fixed route with guide wire let into the floor and travel at a controlled speed. Main precaution is the provision of a trip device fixed to the front of the towing truck such that the truck with full load will stop before the main body of the truck reaches any obstruction it meets.

16.4.5 Road vehicles

Where road vehicles are used solely for transport around a site they should be:

- maintained to *Road Vehicles (Construction and Use) Regulations 1986* standards
- driven in accordance with the Highway Code
- have separate entrances into buildings from pedestrians
- use a banksman or audible warning when reversing.

Site roads should:

- conform to standards for road traffic
- be well defined and separate from pedestrian footpaths
- be maintained in good condition

- be provided with guard rails where pedestrian exits from buildings let straight onto a roadway
- have speed limits [between 15 and 20 mph for medium-sized site]
- have 'sleeping policemen' to restrict speeds.

Essential ingredients for the safe use of powered vehicles on a site are:

- competence of the driver
- the driver's awareness of manoeuvres that could put others at risk
- the driver's taking action to check that the way is clear for the vehicle to move, whether forwards or backwards
- giving priority to pedestrians.

Various types of powered vehicles can be used safely around sites, but the degree of safety is determined by the driver's skill and consideration for others.

Practical guidance on workplace transport is given in HSE publication HSG 136 'Workplace transport safety'.

17 Safe use of electricity

Electrical power is the most convenient and flexible source of power in both industry and in the home. It is also probably the most dangerous because:

• you cannot see it
• you cannot smell it
• if you feel it, it may be too late.

But it can be used safely if its dangers are understood, certain simple rules are followed and it is treated with respect.

Legislation lays down basic requirements to be met in the use of electricity but there are, in addition, a number of well tried techniques that should also be followed. Both these aspects are dealt with in this chapter.

17.1 The Electricity at Work Regulations 1989

The regulations relating to the safe use of electricity at work are relevant to all work situations, this includes mines and quarries where there are particular and specific requirements. This section deals only with the requirements for the safe use of electricity in normal industrial and commercial workplaces.

17.1.1 Legislation

• The *Electricity at Work Regulations 1989* supported by HSE's publication no. HSR 25 'Memorandum of Guidance on the Electricity at Work Regulation 1989'.
• Regulations 14–19 of the *Provision and Use of Work Equipment Regulations 1998* (PUWER).
• Although not legislation, the Institution of Electrical Engineer's Regulations for Electrical Installations (The IEE Wiring Regulations) now incorporated in BS 7611 lays down standards for electrical installations up to 1000 volts ac that are recognized as giving compliance with statutory requirements.

Although made under the HSWA, which requires that standards have to be met *so far as is reasonably practicable*, because of the dangers inherent in the use of electricity certain of the regulations are 'absolute' requirements, i.e. they must be complied with regardless of cost. However, it is recognized that if employers have taken *all reasonable steps and exercised all due diligence to avoid committing an offence* they will have fulfilled this absolute duty, but it will be up to them to prove it if an inspector queries it.

The regulations to be complied with *so far as is reasonably practicable* are the following:

r.2 Defines:

- **conductor** as carrying electrical energy
- **danger** as risk of injury
- **electrical equipment** as anything that uses, generates, transforms, conducts, stores, measures, etc., electrical energy
- **injury** a death or personal injury caused by electricity
- **system** as including all electrical equipment deriving power from a common source.

r.3 Duties are placed on:

- employers and the self-employed to comply
- employees to:
  - co-operate with the employer in complying
  - comply themselves.

r.4 To prevent danger:

- all systems shall:
  - be properly constructed
  - be properly maintained
- any work on a system shall be carried out in a safe manner.

r.6 Electrical equipment working in a hostile environment must be designed and built to resist it.

r.7 All conductors in a system must:

- be insulated and
- be protected or
- have precautions taken to prevent danger.

rr.17–28 Relate to mines and quarries and are not covered by this part.

r.29 Lists those regulations that are absolute but allow a defence of *having taken all reasonable steps and exercised all due diligence*.

Those *absolute* regulations that must be complied with are:

r.4 (part) Protective equipment provided shall be:

- suitable for its intended use
- well maintained
- properly used.

r.5 No electrical equipment to be put into use if overloading its strength and capacity causes danger.

r.8 Where there may be danger from a conductor it shall be earthed or protected by other means.

r.9 Earth conductors must not contain switches or other means that could break the connection.

r.10 Every joint and connection must be sound.

r.11 Excess current protection must be provided to prevent danger.

r.12 Electrical equipment must be provided with:

- means to cut off the supply of electrical energy
- an isolating switch that can be secured (locked) in the OFF state.

Generating equipment is excluded but must take precautions to prevent danger.

r.13 Where equipment has been made dead for working on, precautions must be taken to ensure it remains dead.

r.14 No one may work on a live uninsulated conductor unless:

- it is not reasonable to make it dead
- it is reasonable for the work to be done live
- suitable precautions have been taken, including the provision of protective equipment.

r.15 When working on or near electrical equipment there must be:

- adequate working space
- adequate access
- adequate lighting.

r.16 Where technical knowledge is necessary to prevent danger, the person working must:

- possess that knowledge or
- be experienced or
- be adequately supervised.

In addition PUWER requires that work equipment must:

r.14 • be provided with controls for:
    – starting and re-starting
    – changing operating conditions without danger.

r.15 • permit stopping without danger and, if necessary, isolating power supplies.

r.16 • have emergency stopping which over-rides all other controls.

r.17 Controls must be:

- clearly visible
- identifiable with marking if necessary
- positioned so they can be operated safely
- positioned so operator can see all parts of equipment; if not, controls must include audible warning and delay of start facility.

r.18 Requires that control systems must:

- be safe
- not create danger in operation
- not allow faults to create danger
- accommodate power failures safely
- not interfere with stop and emergency-stop control.

r.19 Controls must include means of isolation from sources of energy which must be:

- clearly identified
- readily accessible (and reachable).

These Regulations lay down basic requirements for the installation and use of electricity on which can be built working practices that enable full and safe use to be made of this source of power.

Guidance on the safe use of electricity is given in HSE publications:

- HSR 25 Memorandum of guidance on the Electricity at Work Regulations 1989
- HSG 47 Avoiding danger from underground services
- HSG 85 Electricity at work. Safe working practices
- HSG 107 Maintaining portable and transportable electrical equipment
- HSG 118 Electrical safety in arc welding
- HSG 141 Electrical safety on construction sites
- HSG 230 Keeping electrical switchgear safe
- HSG 238 Out of control. Why control systems go wrong?
- HSG 253 The safe isolation of plant and equipment
- L 22 Safe use of work equipment
- L128 The use of electricity in mines. *Electricity at Work Regulations 1989*. Approved Code of Practice.

17.2
Safe use of
electricity

Remember – electricity is lethal:

- you can't see it
- you can't smell it
- if you feel it, it may be too late.

When considering the safe use of electricity a number of terms arise that have particular meanings. Those terms include:

- **charged**
  - has acquired an electric charge:
    * through being connected to a live conductor
    * by electrical induction
    * of static electricity
- **competence**
  - having:
    * adequate technical knowledge
    * adequate experience of the work involved
    * a detailed knowledge of the particular process or equipment
  - so as to be able to:
    * recognize defects
    * assess their importance
    * recommend any necessary remedial action
- **dead**
  - carrying no charge, disconnected from all sources of electricity and connected to earth
- **earthing**
  - connecting direct to the mass of the earth so as to prevent any charge building up or being acquired. In earthed circuits the connection from the appliance to earth must be *solid* and not pass through any means of breaking the connection, i.e. switches
- **excess current protection**
  - means to prevent a circuit or appliance being subjected to a current beyond its capacity, usually either a fuse or circuit-breaker set to operate at a predetermined current level
- **double insulation**
  - the provision of two separate layers of insulation between the live parts and the part being handled. Double insulated appliances do not need to have an earth connection
- **duty holder**
  - a person who owes a duty. Under the *Electricity at Work Regulations 1989* this includes anyone who is in control of plant or equipment which, directly or indirectly, can cause danger or injury to other persons
- **insulation**
  - the protection provided on a conductor to prevent it making contact with another conductor, with earth or being touched by a person. The insulation may also have to provide protection against mechanical damage to the conductor

- **IP rating**
  - an international system for classifying levels of protection of equipment against the ingress of dusts and moisture. (IP = Index of Protection – see BS EN 60529 for details)
- **isolation**
  - the switching off of electrical supplies and locking the isolating switch in the OFF position
- **live**
  - being at a voltage through being connected to a source of electricity, i.e. connected to a conductor
- **locking off**
  - a facility on switches whereby they can be locked in the open circuit condition, i.e. locked in the OFF position
- **electrical protection**
  - of plant and equipment:
    * by the provision of fuses or circuit breakers
  - of persons by the provision of:
    * insulation
    * 110-volt centre tapped to earth supply
    * residual current device (RCD) in 240-volt supply
    * means to discharge static electricity charges
- **reduced voltage**
  - use of a voltage below the supply level for ancillary services and to protect operators. Typically, reduced voltage levels are below 50 volts
- **residual current device (RCD)**
  - a device that senses leakage of current to earth and breaks the supply. Nominally senses leakage of 30 mA and breaks the circuit within 30 msecs.

**17.2.1
Dangers from
electricity**

The following dangers are associated with electricity:

- shock:
  - usually from live wire to earth
  - causes muscular spasm
  - can interfere with the regular action of the heart and cause fibrillation of the heart muscles or complete cardiac arrest
  - can cause respiratory failure
- burns:
  - from contact with electric arc caused when conductors short circuit or equipment is overloaded
- arc eyes:
  - from ultraviolet rays when looking at electric arc or welding flash
  - symptoms like conjunctivitis
  - temporary condition lasting three or four days
  - does not affect contact lens
- fire:
  - from electrical arc
  - overloading of conductors
  - discharge of static electricity

- static:
  - caused when two materials are parted, e.g. web from roller; solvent being poured from container
  - high voltage, low current
  - causes spasm of voluntary muscles and violent body movement when injury results from hitting equipment not from the static itself.

17.2.2
Safe use techniques

This part is not concerned with the installation of electrical equipment – that should be done by trained and competent electricians – but with the use of electricity in the workplace.

- Faults:
  - if faults occur, do *not* interfere with electrical equipment – get a qualified electrician
- Circuit protection:
  - the function of a fuse or a circuit breaker is to prevent the down stream circuit from being subjected to current beyond its capacity
  - fuses allow some excess current to flow before they 'blow'
  - circuit breakers trip at the set current
  - all electrical circuits should be protected by a fuse or circuit breaker
  - fuses/circuit breakers should be rated to protect the equipment served
  - fuses/circuit breakers should not be uprated except by qualified electrician
- Isolating switches
  - every piece of equipment using electrical power should have its own isolating switch
  - isolating switches should have means for locking them in the OFF position
  - isolating switches should be labelled to identify equipment served
  - isolating switches must be within easy reach from the working floor level
- Maintenance
  - when working on equipment do not rely on the normal OFF switch – switch off at the isolator and lock it off
  - locking off should be by individual padlock
  - use special multi-padlock calipers if more than one person is working on equipment
  - strict rules should apply to locking-off procedures – see below
- Earthing
  - all electrical equipment, except where double insulated, must be earthed
  - earth circuits must be solid connected to earth and not pass through switches
- Access
  - access space at least 1m wide must be left in front of electrical switch-gear
  - a hand rail positioned 1 m out from switch-gear will keep space clear

- Protection:
  - new installations and existing equipment working in a hostile (wet) environment should have earth leakage protection such as residual current devices (RCDs)
  - all electrically driven machinery should have emergency stop switches which are:
    * readily identified
    * within easy reach of operators
  - interlock switches should be limit switches (*not* micro-switches)
  - actuation of single interlocking switches should be positive, i.e. safety circuit made when switch in the relaxed (non-operated) position
  - all conductors must be insulated – bare wires or exposed connections are *not* permitted
  - for portable equipment '110-volt centre tapped to earth' supply should be used
  - display placards:
    * summarizing the *Electricity at Work Regulations 1989*
    * showing means of artificial respiration (resuscitation).

17.2.3 Rules for locking-off	
1	All electrical machinery should have isolators with locking-off facilities.
2	Before starting any maintenance, repair or other work that requires access into the machinery the isolator must be locked-off by padlock and identifying tag attached.
3	Each padlock should have only one key. There should be no duplicates or master keys
4	Only the person who attached the padlock may remove it. Arrangements may need to be made to transfer padlocks (or keys) at shift change-overs.
5	If more than one person is working on the equipment, multi-padlock hasps should be used and each person attach their own padlock.
6	On major maintenance a single padlock can be used for a gang, in which case the supervisor/foreman carries the responsibility for the safety of the whole gang and for ensuring they are all clear of the equipment before removing the padlock.
7	Before removing the last padlock on completion of the work, the equipment must be checked to ensure all tools have been removed, guards replaced and the equipment is safe to operate.
8	Padlocks should either be issued on a personal basis or kept centrally and signed for at each use.
9	Loss of padlock keys should be reported to the supervisor and the written authority of a responsible manager obtained before the padlock is removed forcibly.
10	In an emergency, if the 'owner' of a padlock is not available, the authority of a responsible manager should be obtained before a padlock is removed.

11 Any employee leaving a padlock on an isolator unnecessarily at the end of a shift should be brought back to work to remove it.
12 Breach of these rules should be subject to disciplinary action.

By following these various proved safe practices and rules full and safe benefit can be obtained from the use of electricity.

**17.3
Safe use of
portable electrical
equipment**

This section concerns any piece of electrical equipment that is carried about during its normal use and which uses a mains power supply. Typical equipment includes hand lamps, any hand-held power tools, portable pipe-threading machines, fans, etc.

17.3.1
Precautions

For various types of portable electrical equipment the following precautions should be taken:

- Supply:
  - 110-volt centre tapped to earth:
    * can be an installed supply or
    * from 240-volt supply via a suitable portable transformer.
  - if 240-volt supply a residual current device (RCD) must be used
  - in both cases the supply must include an earth connection
- Equipment:
  - be either:
    * earthed or
    * double-insulated
- Plug:
  - suitable for the supply outlet
  - properly wired including clamping of the cable sheath
  - the earth wire should have plenty of slack in the plug so it is the last wire to pull out of the terminals thus retaining earthing integrity
  - in good condition
  - fitted with the correct fuse
- Cable:
  - of suitable capacity for the appliance, both voltage and current
  - of flexible type
  - in good condition without any damage to the sheath
  - have earth conductor except in case of double-insulated equipment
  - inspected regularly
- Appliance:
  - in good condition and repair
  - properly wired
  - suitable for the supply voltage
  - the sheath of the supply cable securely clamped
  - properly earthed unless double-insulated
- Inspection:
  - all portable powered equipment (plug, cable and appliance) must be inspected regularly and have suitable dated tag attached.

**17.3.2
Likely faults**

The following faults can occur and are relatively easy to fix:

- damaged cable:
  - replace or insert proper connector at point of damage
- grommet missing at point of entry to appliance:
  - replace grommet
- sheath pulled out of clamp, i.e. the individual wires can be seen:
  - replace and clamp securely on the sheath
- broken plug:
  - replace.

These precautions are simple and easy to take but they do provide a high level of protection in the use of portable equipment. They are as applicable at home as they are at work.

Advice is given in HSE publication HSG 107: 'Maintaining portable and transportable electrical equipment'.

18   Fire

After accidents, fire is one of the largest drains on industrial resources. Not only is it a drain but it is also a major cause of companies going out of business. A fire can remove overnight the manufacturing ability of a company and without that ability customers are lost. No customers, no business. So the prevention of fire can play an important role in the continuing viability of an organization.

The general powers of fire authorities to fight fires are contained in *The Fire and Rescue Services Act 2004* which requires the fire brigade to operate in an efficient and organized manner, to ensure there is an adequate supply of water for fighting fires and gives them the right of entry to buildings if a fire is suspected. It also outlines the reduced role of the fire and rescue authority in employment fire prevention matters.

With the coming into effect of *The Regulatory Reform (Fire Safety) Order 2005* much greater emphasis has been put on the role of the employer in matters of fire precautions and prevention in the workplace. This Order repealed the general fire duties imposed on sports grounds by *The Fire Safety and Safety of Places of Sport Act 1987* but left in place specific duties in respect of the safety of the grounds. The Order also revoked *The Fire Precautions (Workplace) Regulations 1997*.

18.1 Fire and rescue authorities

Set up under *The Fire and Rescue Services Act 2004*, the fire and rescue authorities, which are the successors to the Fire Brigade, have a reduced role in employment fire prevention and an enhanced role in rescue following a road traffic and other accidents.

In respect of places of work they no longer have the responsibility for issuing Fire Certificates but are required to:

- promote fire safety
- give advice on fire precautions, fire prevention measures and means of escape
- provide an effective fire fighting service
- ensure their members are adequately trained and competent
- investigate the causes of fires.

To assist them in carrying out these duties, they have been given powers to:

- enter buildings if they suspect a fire
- ensure an adequate supply of water for fire fighting.

**18.2
Fire safety**

The requirements concerning fire safety in places of work are contained in *The Regulatory Reform (Fire Safety) Order 2005* which puts the responsibility for implementing fire prevention and precautions firmly with the responsible person, be he employer, tenant or landlord. The Order requires in:

Part 1   General

r.2   Defines, inter alia:

- **fire inspectors** as inspectors appointed under s.28 of *The Fire and Rescue Services Act 2004* to check the effectiveness of the fire service
- **inspectors** as inspectors appointed to enforce this Order
- **owner** as any person receiving the rack rent
- **relevant person** as any person who is or maybe on the premises or in the vicinity who could be at risk if there was a fire.

r.3   Defines the responsible person as:

- the employer or
- the person in control of the premises , including the tenant, or
- the owner.

r.4   Lists general fire precautions as measures:

- to reduce the risk of fire
- relating to the means of escape
- to fight fires
- to detect and warn of fires
- concerning the provision of fire training and escape.

r.5   Places duties on:

- the employer
- the person in control of premises
- tenants.

r.5   Places duties on the responsible person to comply with the requirements of this Order

r.6   Excludes:

- domestic premises
- offshore installations
- ship board activities
- agriculture and forestry away from buildings
- road and rail transport
- mines.

r.7   Excludes certain requirements for;

- occasional working
- a family undertaking
- air, land and water transport
- armed forces, security and emergency services and prisons.

Part 2 Fire safety duties

r.8 Requires employers to take precautions to protect their employees and others likely to be affected by fires

r.9 Requires the responsible person to:

- carry out a fire risk assessment covering items listed in Schedule 1
- review the assessment
- not employ a young person until an assessment of particular risks made
- record the fire risk assessment in writing where:
  - five or more employees
  - a statutory licence has been issued for the premises
  - an alterations notice has been issued
- not start new work with a dangerous substance until a risk assessment made.

r.10 In implementing fire precautions, employers should follow the principles;

- avoid risks
- evaluate risks that cannot be avoided
- combat risks at source
- adapt to technical progress
- replace the dangerous with non- or less-dangerous
- develop an overall coherent prevention policy
- give collective measures priority over personal protection
- instruct employees.

r.11 The responsible person must have in place fire safety arrangements appropriate to his organization. These must be recorded.

r.12 The risks from dangerous substances must be reduced so far as is reasonably practicable following the principles of prevention (Schedule 1 part 3)

r.13 Premises must have:

- fire fighting equipment
- fire detectors
- fire alarms
- portable extinguishers
- competent fire warden
- contact with the local emergency services.

r.14 Emergency escape routes must:

- lead directly to a place of safety
- allow speedy evacuation
- be sufficient for the number working
- have doors that open outwards (sliding doors not allowed)

- have escape doors that can be opened easily
- be indicated by suitable signs
- have emergency lighting where necessary.

r.15 In cases of imminent and serious dangers there should be:

- escape procedures with fire drill practices
- trained nominated persons to ensure safe escape
- adequate information on hazards and precautions
- no work until danger eliminated.

r.16 Where sufficient quantities of dangerous substances are present there must be:

- information on:
  - the emergency arrangements
  - hazards and their identification
  - specific hazards arising from an incident
- suitable means for warning
- special escape facilities
- information for the emergency services of hazards
- hazard warning signs displayed
- in the event of a fire:
  - action must be taken to mitigate the effects
  - inform those affected
  - allow only nominated persons to carry out repairs.

r.17 Maintain the premises in efficient working order and good repair

r.18 Appoint a competent safety adviser, preferably a qualified employee

r.19 Provide information on:

- risks identified by fire risk assessments
- precautions and preventative measures
- procedures in an emergency
- identity of competent advisers
- fire risks notified by joint tenants
- fire risks and precautions to parents of child employees
- dangerous substances used.

r.20 Provide like information to contractors and employers of temporary workers

r.21 Provide suitable fire safety training:

- when first employed
- when job or work system changes
- periodically.

r.22 Co-operate with joint tenants and other users of the premises

r.23 Places duties on employees to:

- take care of themselves (HSWA cl. 7 (a))

- co-operate with employer in complying
- inform employer of:
  - any risks identified
  - shortcomings in employer's arrangements.

r.24 Gives power to the Secretary of State to make regulations

Part 3   Enforcement

r.25 Enforcing authorities are:
- Fire and Rescue Authority except:
  - HSE for:
    * nuclear sites
    * ships
    * construction sites
  - fire service of the Ministry of Defence for armed services premises
  - local authority for sports grounds
  - fire inspector for Crown premises.

r.26 Above authorities to enforce this Order except that the fire and rescue services may arrange for HSE or Rail Regulator to act on its behalf

r.27 Inspectors appointed by the enforcing authorities must carry a warrant or other certificate of authority and produce it on demand. They have powers to:

- enter any premises at any time without force
- make any necessary enquiries
- require the production of statutorily required records
- require occupier to provide facilities to carry out inspection
- take samples of articles or substances
- require high-fire hazard articles and substance to be destroyed or dismantled
- consult with employers about fire risks on the premises.

r.28 • delegate powers to officers of the Fire Brigade

r.29 • Enforcing authorities can issue *Alteration Notices* where fire safety could be compromised by:
  - condition of the premises or
  - changes to or in the premises
- *Alteration Notices* stand until withdrawn or cancelled by a court.

r.30 • Issue *Enforcement Notices* for failure to comply with this Order

r.31 • Issue a *Prohibition Notice* for serious imminent risk of fire

Part 4   Offence and appeals

r.32 Non-compliance with this Order is an offence that attracts a variety of penalties r.33 Defences include *took all reasonable precautions* and *exercised all due diligence* or

r.34 it was not reasonably practicable to comply

r.35 Appeals against a Notice to be made to a magistrate's court within 21 days of service of notice

r.36 Disputes and differences referred to Secretary of State for resolution

r.37 & 38 refer to protection of fire fighters

r.39 No rights to civil action for breaches of this Order except for damage to an employee

r.40 Not charge employees for anything done under this Order

r.41 Employers have a duty to consult employees

r.42 Refers to special conditions for premises licensed under this Order

r.45 Requires the local authority to consult the enforcing authority before passing planning applications

---

**18.3
Causes of fire
and precautions**

Fires don't just happen, they are caused when three elements are present. These are:

1 Oxygen

- normally from the air which contains 20% oxygen
- can be given off by oxidizing chemicals such as nitrate fertilizers.

2 Fuel

- can be anything combustible:
  - with solids the smaller the pieces the more easily it ignites
  - with liquids, the lower its flash point the more readily it ignites
  - with gases the concentration needs to be within its flammability limits.

3 Ignition

- caused by any source that will raise the temperature above the flash or ignition point, including:
  - cigarette ends
  - electrical sparks and short circuits
  - static electricity

     – heated equipment and over heated bearings
     – heating pipes
     – sparks from welding and burning operations.

All three must be present for a fire to start. Remove any one element and the fire dies. This is shown diagrammatically in Figure 18.1, the fire triangle.

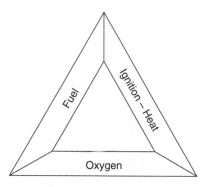

Figure 18.1 The fire triangle

Causes of fires	Precautions
Bad housekeeping	– regular cleaning with rubbish disposed of into suitable receptacles
Electrical	– do not overload circuits, ensure fuses/circuit breakers operate at correct current level – provide adequate cooling air flow – do not allow fluff to build up on electrical equipment – ensure electrical contacts are kept clean
Hot bearings	– keep clean and well maintained
Heating pipes	– ensure all combustible material is stored well clear of pipes
Welding and gas cutting	– remove combustible material from area – have suitable extinguishers positioned nearby – post fireman during actual work and for half-hour afterwards
Smoking	– prohibit
Oils and solvents	– store outside – use only enough for day or shift – use non-spill containers for handling – use earth link when pouring
Oily rags	– dispose of in non-flammable (metallic) bins.

18.3.1 Heat transfer	Heat transfer can be by: 1 **Conduction** – the carrying of heat along or through materials 2 **Convection** – heat carried by rising currents of hot air 3 **Radiation** – the emission of long wavelength infra-red rays.
18.3.2 Fire spread	Fires can spread easily:

- across open plan layout:
  - normally spreads at high level therefore provide high-level ventilators/extractors to draw flames out of building
  - divide up area into fire compartments
- through gaps or holes in fire-break barriers
  - ensure all gaps in fire-break barriers are plugged with fire resisting material particularly round pipe and cable entries
  - ensure fire-break doors are kept shut or have means for closing them automatically if fire occurs
- along channels and gullies in which flammable liquids and heavier-than air gases can flow:
  - for liquid spills:
    * bund round container
    * use suitable absorbent granules
  - for heavier-than-air gases and fumes ensure the area is well ventilated at low level
- across the surface of dusts in roof spaces
  - keep roof spaces cleaned
  - build fire breaks in roof space
- carried through ventilating and exhaust ducting
  - switch off ventilating fans
  - provide fire extinguishant in the ducts
- along corridors
  - fit smoke/fire doors and keep them shut or fit magnetic releases linked to fire-alarm system
- in lift shafts
  - shaft well should be enclosed
  - fit smoke/fire doors on landings at access to work areas and corridors.

The main thrust of fire precautions is to save life by preventing a fire from starting in the first place, but if a fire does occur the next priority is to ensure the layout of the building is such that it will ensure the escape of the occupants and will stop the fire from spreading and hence keep damage and disruption to a minimum.

18.4 **Fire-fighting and** **extinguishers**	This deals only with 'first-aid' fire-fighting and not the fighting of major conflagrations. Most fires start out very small and can be safely tackled at that stage by means of hand-held extinguishers. However, at no stage should employees put themselves at risk in attempting to put a fire

out. They should always place themselves between the fire and the escape route.

<table>
<tr><td>

**18.4.1
Fire-fighting
equipment**

</td><td>

The types and number of equipment are determined by:

- fire risk assesment
- advice of the local Fire Prevention Officer (for which you may have to pay)
- advice from fire insurer's surveyor
- pressure from a fire extinguisher salesman (treat their recommendations on numbers and the need for replacement units with suspicion – always check recommendations with the Fire Prevention Officer before buying).

</td></tr>
<tr><td>

**18.4.2
Types of equipment**

</td><td>

There are two types of equipment: installed and hand-held.

1 Installed equipment

- hose reels:
  - need reliable water supply
  - suitable number to cover whole work area
  - can be a nuisance if stop cock leaks
  - can be arranged to turn water on as hose is pulled out
- sprinklers:
  - usually in high-risk areas
  - liked by fire insurers who reduce premiums
  - each sprinkler head spray covers approx. 10 m^2 (100 ft^2)
  - will contain the fire but can cause a great deal of damage to stock
  - need guaranteed supply of water at pressure
  - expensive to install and maintain
- halogen gas:
  - in computer and electrical control gear
  - suppresses fire but does not remove heat so fire can restart when gas turned off or runs out
  - being replaced by water in some computer installations
- carbon dioxide:
  - in electrical substations
  - for solvents on printing machines
  - is an asphyxiant so all employees must be clear before gas turned on
  - does not remove the heat so the fire can restart if the gas runs out or is turned off.

2 Hand-held extinguishers

There are six basic types of these whose containers are painted red but with a colour coded band or panel to identify its contents and the sort of fire for which it is suitable.

</td></tr>
</table>

Type	Colour of panel	Suitable for
Water	red	carbonaceous materials – wood, paper, coal, etc.
Foam, including aqueous film forming foam (AFFF)	cream	carbonaceous materials, liquids whether soluble in water or not
Carbon dioxide	black	liquids whether soluble in water or not; fires in electrical equipment
Dry powder	blue	liquids whether soluble in water or not; fires in electrical equipment
Vaporizing liquid	green	liquids whether soluble in water or not; fires in electrical equipment
Wet chemical	canary yellow	cooking oils and fats

EU Directive requires that all extinguishers supplied after May 1997 must be painted red but allows 5% of body to be a different colour. Existing extinguishers need not be repainted.

18.4.3 Classes of fires

Class	Materials	Extinguishant
A	Organic materials – wood, coal paper	Water, AFFF
B(i)	Liquids and liquefiable solids soluble in water – acetone	Foam, vaporizing liquids, CO_2, dry powder, AFFF
B(ii)	Liquids and liquefiable solids not soluble in water – petrol, fats, wax	Foam, AFFF, vaporizing liquids, CO_2, dry powder
C	Gases and liquefied gases – propane, butane	Isolate supply, drypowder
D	Metals – magnesium, aluminium	Special dry powder
F	Cooking oils and fats	Wet chemical

Electricity can cause a fire in any class – use only vaporizing liquid, dry powder or CO_2 extinguishers.

Fire-fighting equipment should be:

- mounted on red-painted fire points
- well identified
- located adjacent to fire exit
- can include an alarm point.

Training to be provided to:

- all staff in fire drill
  - what to do when alarm sounds
  - assembly points
  - roll call
  - role of fire marshals
- supervisors and selected staff:
  - in checking that allocated areas of the workplace have been evacuated
  - in use of extinguishers
- volunteer fire teams:
  - in first-aid fire-fighting.

Action on finding a fire:

1 Sound the alarm (break glass).
2 Call fire brigade.
3 Use extinguisher if safe to do so.
4 Evacuate premises by quickest route.
5 Go to assembly point for roll call.

Where's the fire then?

Like oak trees, fires start from small beginnings. If they can be tackled at that stage an enormous amount of disruption and loss can be prevented. However, in no circumstances should employees put themselves at risk in tackling even a small outbreak.

**18.5
Safe use of
flammable
substances**

Many highly flammable substances are used at work and in spite of their flammable nature can be used quite safely if certain precautions are taken.

Requirements for their safe use and storage were contained in the *Highly Flammable Liquids and Liquified Petroleum Gases Regulations 1972* which have been revoked by the *Dangerous Substances and Explosive Atmospheres Regulations 2002*. However, they did lay down sound precautions to be taken where highly flammable materials are used and potentially explosive atmospheres arise. Their contents are summarized below:

Liquefied petroleum gases (LPG)

- butane and propane
  - stored in bulk or in portable cylinders
  - cylinders to be:
    * stored out of direct sunlight
    * kept in well ventilated area
    * fitted with pressure relief valve
  - after use turn off at valve on cylinder
  - do not rely on the integrity of hose to prevent leakage
  - check hose regularly for signs of cracking and splitting
  - use in well ventilated areas
  - no smoking in area where LPG in use or store.

Welding and burning gases

- oxygen, hydrogen,
  - stored in high pressure cylinders
- acetylene
  - stored by dissolving in absorbent medium in high pressure cylinder
- all high pressure flammable gases
  - store cylinders of gases separately, i.e. do not mix cylinders of different gases in storage
  - store cylinders out of direct sunlight
  - store cylinders upright
  - in use restrain cylinders upright in proper frame
  - use anti-flashback fitting in gas line
  - check gas hoses regularly for leaks or damage.

Flammable liquids

- petrol, paraffin, white spirit, etc.
  - store outside building
  - bund the storage area
  - use earth link when transferring liquids
  - use drip tray in decanting area
  - transport in non-spill containers
  - use only in well ventilated areas
  - ban smoking in area
  - operators to wear protective gloves.

Advice on the safe use and storage of flammable liquids is given in HSE publications:

HSG 51: The storage of flammable liquids in containers
HSG 176: The storage of flammable liquids in tanks
HSG 178: The spraying of flammable liquids
HSG 186: The bulk transfer of dangerous liquids and gases between ship and shore.

Oxygen enrichment

- occurs when using oxgen in confined or restricted spaces:
  - ensure space is well ventilated
  - turn oxygen supply off at cylinder or manifold valve when not being used – even for short tea or meal breaks.

Foam plastics

- store bulk away from work area
- have only supplies for day or shift in work area
- enforce no-smoking
- keep escape routes clear of material
- dispose of scraps in suitable containers.

Newly introduced regulations *The Dangerous Substances and Explosive Atmospheres Regulations 2002* (DSEAR) deal with those substances – whether solids, dusts, liquids, vapours or gases – that, when present in the atmosphere, can ignite and cause injury or damage by explosion, fire or 'other event'. The Regulations apply to all premises where such circumstances can occur but with certain listed exceptions.

In these Regulations:

r.2   Defines

- dangerous substances as:
  - a substance classified as explosive, oxidizing, extremely flammable, highly flammable or flammable in CHIP 2002
  - other substances which because of their physico-chemical or chemical properties or the manner in which they were used, created a risk
  - any dust which can form an explosive mixture with air
- explosive atmosphere as a mixture of air and one or more dangerous substances that can be ignited and will support combustion.

r.3   Excludes, inter alia, ship board activities, medical treatments, gas used for cooking or heating, activities in mines and quarries.

Duties are placed on employers:

r.4   • extending protection to non-employees affected by the work

r.5   • requiring the carrying out of a risk assessment when danger-
        ous substances are present and an explosive atmosphere may
        develop

r.6   • requiring the elimination or reduction of risks as far as possi-
        ble, advocating substitution of less dangerous substances
      • giving an order of priority for controlling risks:
        – reducing the quantity present to a minimum
        – avoiding release or controlling release at source
        – preventing explosive atmospheres by ventilation
        – containing or collecting any leakage
        – avoiding sources of ignition or conditions that could cause
          harmful effects
        – segregating incompatible substances
        – reducing the number of employees exposed
        – incorporating explosion relief in plant design
        – incorporating explosion suppression devices in processes
        – design plant to withstand likely explosions
        – providing suitable PPE
      • ensuring the safe handling, storage and use of dangerous
        substances
      • avoiding conditions that could create danger
      • taking precautions in the design of plant and work processes
        or by organizational arrangements.

r.7   Areas with a potential for explosive atmospheres to be:

      • identified
      • classified into zones:
        – zone 0 continual risk of explosion
        – zone 1 explosion risk likely in normal operation
        – zone 2 explosion risk but not likely in normal operation but
          could occur occasionally
        – zone 20 where clouds of combustible dust are present for
          long periods or frequently
        – zone 21 where clouds of combustible dust occur during
          normal operations
        – zone 22 where clouds of combustible dust are present only
          for short periods
      • equipment used in these zones should be:
        – zones 0 and 20 category 1 equipment
        – zones 1 and 21 categories 1 and 2 equipment
        – zones 2 and 22 categories 1, 2 and 3 equipment

[Categories of equipment are given in *The Equipment and Protective Systems intended for use in Potentially Explosive Atmospheres Regulations 1996*]
- PPE must not generate electrostatic discharges.

r.8 Unless the risk assessment shows the risk to be small and the actions listed in r.6 have been taken, the employer must have in place arrangements for dealing with emergencies, including:

- first aid facilities and safety drills
- information on emergency arrangements
- visual and audible warning of impending emergency
- communication with the emergency services
- maintenance of escape facilities
- steps to mitigate the effects of the emergency and facilitate the return to work
- suitable PPE and any necessary specialized equipment.

r.9 Employees must be provided with:

- information, instruction and training in the precautions to take
- name and details of the dangerous substances
- findings of the risk assessment.

r.10 Containers and pipes carrying dangerous substances to be marked with the contents.

r.11 Employers sharing premises to co-operate in implementing precautionary arrangements.

r.13 Lists the revocations including:

- Section 31 of the Factories Act
- *The Highly Flammable Liquids and Liquified Petroleum Gases Regulations 1972.*

A good working environment has a great influence on the health of the employees and on their attitude towards their work. Also, the atmosphere within the workplace and how it gets rid of harmful fumes has a large effect on the local community. *The Control of Substances Hazardous to Health Regulations 2002* (COSHH) were aimed at providing protection for the health of employees inside the workplace and the *Environmental Protection Act 1990* (EPA) set out to protect the community from a whole range of potential health hazards including those caused by work activities. The method of disposal of hazardous wastes is regulated by *The Hazardous Waste (England and Wales) Regulations 2005* and *The Waste Management (England and Wales) Regulations 2006.*

COSHH is dealt with in Section 11.1. EPA and *The Hazardous Waste Regulations* are dealt below.

19.1 The Environment Protection Act 1990

This Act set out to deal with the whole question of pollution of the environment and is aimed at regulating pollution from industrial, domestic and community sources. This section is concerned only with pollution resulting from work activities.

The Act:

Defines:

- **pollutants** as:
  - solid wastes for discharge onto land
  - liquid wastes whether discharged onto land or into waterways
  - discharges into the atmosphere
  - noise in the community
- **controlled waste** as:
  - household waste
  - industrial waste
  - commercial waste
- **special waste** as:
  - controlled waste that is so dangerous that it requires special disposal procedures, i.e. dangerous to life or liquids with a flash point of 21 °C or less.

Has a three-pronged approach:

1   Air Pollution Control (APC):

   – deals with discharges into the atmosphere
   – enforced by local authority officers

2   Integrated Pollution Control (IPC):

   – deals with discharges to:
     * the air
     * the ground
     * waterways
   – enforced by Environment Agency Inspectors

3   Control of discharges onto land.

Gives Minister powers to:

• specify control standards to be met
• 'prescribe' processes requiring authorization
• issue authorization for prescribed processes
• place conditions on authorizations requiring the use of *best available techniques not entailing excessive cost* (BATNEEC)
• charge for issuing an authorization.

Appeals concerning the issue, revocation or alteration of authorizations heard by the Secretary of State.

No one allowed to carry on a 'prescribed process' without due authorization.

Enforcement by inspectors of:

• the Environment Agency (EA) appointed by the Minister
• river purification authority
• local authorities.

Powers of inspectors:

• entry into premises where prescribed processes are carried out
• issue *Enforcement Notices* (similar to Improvement Notices under HSWA)
• issue *Prohibition Notices* (again similar to Prohibition Notices under HSWA)
• carry out investigations
• measure and take photographs
• take samples
• require person to give information
• examine records
• obtain assistance with investigation
• if imminent danger, to render substances harmless.

Enforcing authorities to:

- keep register of prescribed processes and authorizations
- make it available for inspection by members of the public.

Appeal against a Notice is to the Secretary of State.

In a prosecution, a defence for accused is to prove he used BATNEEC.

A *waste management licence* is required for keeping, treating or disposing of controlled waste on land.

Local authority officers inspect for *statutory nuisances* including:

- discharges into the atmosphere
- noise emitted from a premises that is prejudicial to health or a nuisance (see Sections 12.3 and 12.4).

Maintaining a good environment at work is sound commercial sense and has many benefits for the workforce and for the local community.

19.2 Hazardous waste regulations

The legislation concerning the disposal of hazardous waste (previously known as *special waste*) is *The Hazardous Waste (England and Wales) Regulations 2005*. Scotland and Ireland retain the core of *The Special Waste Regulations 1996* but with regional amendments. The aim of the regulations is to ensure, throughout the whole disposal process, the safe disposal of wastes that have proved hazardous to health if they enter the human consumption chain.

These Regulations that bring into UK law the content of the EU Waste Directive (75/442/EEC as amended) are much more extensive and detailed than the *Special Waste Regulations* they replace, but they retain much of its procedural paperwork.

The 2005 Regulations are very comprehensive and are split into twelve parts, each dealing with a specific facet of hazardous waste disposal. Salient points include:

Part 1   General

r.5 Defines:

- Consignor as the person who arranges for the waste to be moved from the premises where it was produced
- Carrier as the person who transports it from the consignor to the consignee

- Consignee as the person who receives the waste for recovery or disposal
- Agency as the Environment Agency in England.

Part 2   Hazardous waste

r.6   Defines hazardous waste as any substance listed in *The List of Waste (England) Regulations 2005*

Part 5   Notification of premises

r.21 Requires the notification to the Agency of premises where hazardous waste is produced and prohibits the removal of such waste until the premises are registered. Registered premises to be given a *premises code*

Part 6   Movement of hazardous waste

r.34 Where hazardous waste is to be consigned it must be identified by a unique consignment code issued by the Agency

r.36 Outlines the standard consignment procedure based on a five-part consignment note which must include the consignment code. The procedure is:

1   the waste producer (consignor) completes parts A and B of copies for himself, consignor, carrier and consignee
2   waste producer gives all copies to the carrier
3   carrier completes part C of all copies and returns them to the consignor
4   consignor completes part D of all copies and retains one
5   consignor gives remaining copies to carrier
6   carrier ensures all copies travel with the waste and gives them to the consignee on delivery of the waste
7   consignee completes part E of all copies and:
    – gives one copy to carrier to retain
    – retains one copy.

The parts of the consignment note are:

A – notification details
B – description of waste
C – carrier's certificate
D – consignor's certificate
E – consignee's certificate.

Schedule 4 of the Regulations contains examples of these consignment notes.

Part 7   Records and returns

R.48 Consignees to keep a detailed record of all hazardous waste disposed of and its location for 3 years or until licence terminated

r.49 Copies of consignment notes to be kept in a register by:

- consignor for 3 years
- carrier for 12 months
- consignee for 3 years or until his permit is surrendered or revoked when register is given to the Agency.

Part 10   Enforcement

r.64 The Agency is responsible for enforcing these Regulations

r.66 Defences allowed in the event of a charge for a breach include:

- took all reasonable precautions
- exercised all due care and diligence
- in an emergency, not practicable to comply but took action to minimize the threat and restore to normal as soon as was practicable.

Schedules 1and 2 cite types of hazardous waste listed in annexes to the Waste Directive

Schedule 3 lists the properties that make waste hazardous (similar to those covered by CHIP)

The List of Waste (England) Regulations 2005 classifies and codes wastes identified as hazardous and gives each a unique reference number. Inclusion of a substance in the List automatically classifies it as hazardous. The List uses the criteria of the *Chemical (Hazard Information and Packaging for Supply) Regulations 2002* when determining whether a substance is hazardous.

Because dangerous chemicals dumped randomly can very quickly leech into a source of drinking water, particular responsibilities are placed on waste site managers through the *Waste Management Licensing Regulations 1994 as amended* and on everyone in the waste disposal chain through the *Environmental Protection (Duty of Care) Regulations 1991.*

Other less hazardous waste referred to in EPA as 'controlled waste' is defined in the *Controlled Waste Regulations 1992* as domestic and commercial waste with some industrial waste.

**19.3
Safe and healthy
working
environment**

A large portion of a person's life is spent in the work environment. A good environment will ensure they stay healthy in both body and mind and be able to enjoy a high quality of life. This section considers some of the factors in the work environment.

- Atmosphere:
  - workplace must be provided with sufficient quantity of fresh or purified air
  - should be free from contaminants such as dusts and fumes [see Section 11.6]
  - extract dust and fumes at source and filter before discharging outside the building
  - have good natural ventilation wherever possible
  - openable windows
  - if forced ventilation (air conditioning):
    * ensure no draughts from outlets
    * check noise levels
    * provide local controls
    * check system for legionella
  - with free-standing fans:
    * ensure blades are securely guarded
    * ensure they do not cause draughts
    * protect trailing cables.
- Lighting:
  - to be bright enough to enable work to be done without strain to eyes
  - walkways adequately lit
  - local lighting for precise work
  - good overall general lighting
  - no points of light likely to cause glare
  - use natural light wherever possible
  - provide blinds or shade from bright sunlight.
  - see CIBSE 'Code for Interior Lighting'
- Cleanliness:
  - work area should be cleaned regularly
  - rubbish disposed of in proper receptacles.
- Overcrowding:
  - ensure each person has at least 11 m^3 (400 cu ft)
  - allow for space taken up by large pieces of equipment
  - provide adequate gangways between work stations.
- Temperature:
  - is required to be *reasonable*
  - not specified but normally taken to be a minimum:
      for sedentary work a minimum of 16 °C (60.8 °F)
      for hard physical work a minimum of 13 8C (55.4 °F)
  - a suitable number of thermometers should be positioned about the workplace.
- Noise:
  - should not be excessive:
  * in manufacturing areas less than 80 dB(A) (see Section 12.1)
  * in offices, laboratories, libraries, etc. not in excess of 40 dB(A).

A clean and healthy working environment is good business sense reducing likely illnesses (and corresponding absences) to a minimum and providing an atmosphere that encourages the workpeople to give of their best.

Appendix 1　Abbreviations

Abbreviation	Meaning
ACAS	Advisory, Conciliation and Arbitration Service
ACOP	Approve Code of Practice
ACTS	Advisory Committee on Toxic Substances
ADR	European Agreement concerning the International Carriage of Dangerous Goods by Road
AGV	Automatic guided vehicle
AIDS	Acquired immune deficiency syndrome
AFFF	Aqueous film forming foam
All ER	All England (Law) Report
APC	Air pollution control
BATNEEC	Best available technique not entailing excessive costs
BITA	British Industrial Truck Association
BOHS	British Occupational Hygiene Society
BS (xx)	British Standard (number)
BSC	British Safety Council
BSI	British Standards Institution
BSS	British Standard Specification
CAS number	The number assigned to a substance by the Chemical Abstract Service
CAW(R)	The Control of Asbestos at Work Regulations 2002
CBI	Confederation of British Industries
CDM	The Construction (Design and Management) Regulations 2007
CE	The mark stamped on machinery and equipment to show that it conforms with EU directives
CEN	Comité Européen de Normalization – European Committee for Standardisation (for mechanical equipment)
CENELEC	Comité Européen de Normalization Electrotechnique – European Committee for Electrical standardization
CHASE	Complete Health and Safety Evaluation
CHIP	The Chemical (Hazard Information and Packaging for Supply) Regulations 2002
CIBSE	The Chartered Institution of Building Services Engineers
CITB	Construction Industry Training Board
CLAW	The Control of Lead at Work Regulations 2002
COSHH	The Control of Substances Hazardous to Health Regulations 2002

CP	Common position (adopted by European Council)
CS	Chemical series (of Guidance Notes)
CSA	Chemical safety assessment
dB(A)	'A' weighted decibel
DBERR	Department of Business Enterprise and Regulatory Reform (ex DTI)
DCLG	Department of Communities and Local Government
DG	Director(ate) General
DSE	Display screen equipment
DSEAR	The Dangerous Substances and Explosive Atmospheres Regulations 2002
DSS	Department of Social Security
DTI	Department of Trade and Industry (renamed DBERR)
EA	Environment Agency
ECA	European Chemical Agency
EcoSoc	Economic and Social Committee of the European Union
EEA	European Economic Association
EEC	European Economic Community (now European Union)
EFTA	European Free Trade Association
EH	Environmental health series (of Guidance Notes)
EHO	Environmental Health Officer
EINECS	European Inventory of Existing Commercial Substances
EL	Employer's Liability (insurance)
ELINCS	European List of Notified Chemical Substances
EMAS	Employment Medical Advisory Service
Employment NTO	National Training Organisation for Employment
EN	European Normalization (prefixed letters to harmonized European standards)
EP	European Parliament
EPA	Environmental Protection Act 1990
ERA	Employment Rights Act 1996
ESR	Essential safety requirements
EU	European Union
FA	Factories Act 1961
FMEA	Failure mode and effect analysis
FTA	Fault tree analysis
GS	General series (of Guidance Notes)
HAVS	Hand and arm vibration syndrome
HAZAN	Hazard analysis
HAZCHEM	Code of hazardous chemicals
HAZOPS	Hazard and operability study
HPA-RPD	Radiation Protection Division of the Health Protection Agency
HSC	Health and Safety Commission
HSE	Health and Safety Executive
HSG	Health and Safety Guidance booklets

HSR	Health and Safety Regulations booklets (being superceded by the L series booklets)
HSWA	The Health and Safety at Work, etc, Act 1974
Hz	hertz (cycles per second)
IAC	Industry Advisory Committee
IBC	Intermediate bulk container
ICE	Internal combustion engine
IEC	International Electrotechnical Commission (for International electrical standards)
IEE	The Institution of Electrical Engineers
IOELV	Indicative Occupational Exposure Limit Value
IOSH	Institution of Occupational Safety and Health
IPC	Integrated pollution control
IP rating	Index of Protection (an international rating system – see BS IEC 60529)
ISBN	International standard book numbering
ISO	International Standards Organization (for international mechanical standards)
ISRS	International Safety Rating System
KISS	Keep it short and simple
L	Legal series of booklets
LA	Local Authority (Council)
LEV	Local exhaust ventilation
LOLER	The Lifting Operations and Lifting Equipment Regulations 1998
LPG	Liquified petroleum gas
LR	The Lifts Regulations 1997
MEGC	Multi element gas cylinders
MEL	Maximum exposure limit
MHOR	The Manual Handling Operations Regulations 1992
MHSW	The Management of Health and Safety at Work Regulations 1999
MIIRSM	Member of the International Risk and Safety Managers
MMMP	Machine made mineral fibres
MP	Member of Parliament
MS	Medical series (of Guidance Notes)
NEBOSH	The National Examination Board in Occupational Safety and Health
NONS (R)	Notification of New Substance (Regulations)
NVQ	National Vocational Qualification
OSRP	The Offices, Shops and Railway Premises Act 1963
PAT	Portable appliance testing
PCB	Polychlorinated biphenyl
PCT	Polychlorinated triphenyl
PD	Published Document (by BSI)
PER	The Pressure Equipment Regulations 1999

PL	Public Liability (insurance)
PM	Plant and Machinery series (of Guidance Notes)
PPE	Personal protective equipment
PSSR	The Pressure Systems Safety Regulations 2000
PUWER	The Provision and Use of Work Equipment Regulations 1998
QMV	Qualified majority voting
r.	Regulation
REACH	Regulation concerning the Registration, Evaluation, Authorization and Restriction of Chemicals
RCD	Residual current device
RID	European Agreement concerning the International Carriage of Dangerous Goods by Rail
RoSPA	The Royal Society for the Prevention of Accidents
RPE	Respiratory protection equipment
RSI	Repetitive strain injury
RSP	Registered Safety Practitioner (of IOSH)
s.	Section (of Act)
SDS	Safety data sheet
SME	Small and medium enterprise
SMSR	The Supply of Machinery (Safety) Regulations 1992
SQWG	Social Questions Working Group (of EU)
SVQ	Scottish Vocational Qualification
TREMCARD	Transport emergency card
TSO	The Stationery Office
TUC	Trades Union Congress
TWA	Time weighted average
UK	United Kingdom
UN	United Nations
VCM	Vinyl chloride monomer
VDU	Visual display unit
VWF	Vibration white finger
WATCH	Working Group on the Assessment of Toxic Chemicals
WC	Water closet
WEL	Workplace exposure limit
WHSWR	The Workplace (Health, Safety and Welfare) Regulations 1992
WRULD	Work related upper limb disorders

Appendix 2 List of statutes

Index